DYNAMIC CHARACTERISATION OF
ANALOGUE-TO-DIGITAL CONVERTERS

THE KLUWER INTERNATIONAL SERIES IN ENGINEERING AND COMPUTER SCIENCE

ANALOG CIRCUITS AND SIGNAL PROCESSING
Consulting Editor: Mohammed Ismail. *Ohio State University*

DYNAMIC CHARACTERISATION OF ANALOGUE-TO-DIGITAL CONVERTERS

by

Dominique Dallet

Laboratoire IXL-ENSEIRB,
Bordeaux, France

and

José Machado da Silva

Universidade do Porto,
INESC-Porto, Portugal

 Springer

A C.I.P. Catalogue record for this book is available from the Library of Congress.

ISBN-10 0-387-25902-3 (HB)
ISBN-13 978-0-387-25902-4 (HB)
ISBN-10 0-387-25903-1 (e-book)
ISBN-13 978-0-387-25903-1 (e-book)

Published by Kluwer Academic Publishers,
P.O. Box 17, 3300 AA Dordrecht, The Netherlands.

www.springeronline.com

Printed on acid-free paper

Printed in the Netherlands.

Contents

Preface

The present book is one of the outcomes of the project DYNAD - Methods and Draft Standards for the Dynamic Characterization and Testing of Analogue-to-Digital Converters. This project was held between 1997 and 2000, supported by the European Commission under the Standards, Measurements and Testing Programme, reference SMT4-CT98 2214, within the Framework IV activities. Its consortium comprised the University of Parma - Italy, the École Nationale Superieure d'Electronique, Informatique & Radiocommunications de Bordeaux - France, Thales (former TTM-Thomson CSF) - France, Italtel Spa - Italy, Infineon Technologies-Development Center Villach - Austria, and INESC-Porto - Portugal. Besides the authors of the different chapters of this book, other people contributed with their work to the start and success of the initiative. We acknowledge the efforts of Hubert Pernull, Otto Wiedenbauer, and Andreas Bertl from Infineon, Roberto Scotti from Italtel, Jorge Duarte and José Matos from INESC-Porto, M. Heuber and M. Zirnheld from Thales, and C. Rebai from ENSEIRB.

A state of the art overview of the methods and procedures employed for characterising the dynamic performance behaviour of analogue-to-digital converters using sinusoidal stimuli, is presented in this book. The three classical methods — histogram, sine wave fitting, and spectral analysis — are thoroughly described, and new approaches are proposed to circumvent some of their limitations.

This is a must-have compendium, which can be used by both academics and test professionals, to understand the fundamental mathematics underlining the algorithms of ADC testing, and as a handbook to help the engineer in the most important and critical details for their implementation.

DOMINIQUE DALLET, JOSÉ MACHADO DA SILVA

Contributing Authors

CHIORBOLI, Giovanni graduated in Electronic Engineering from the University of Bologna, Italy, in 1987. Until 1990, he was with the University of Bologna. Since 1990, he has been with the University of Parma, Italy, where he is currently an Associate Professor of Electronic Measurements. His scientific interests are in the field of electronic instrumentation, analogue-to-digital and digital-to-analogue modelling and testing, and electrical characterization of semiconductor devices and materials for microelectronics.

DALLET, Dominique was born in Rochefort/Mer, France, on July 3, 1964. He obtained his PhD degree in Electrical Engineering in 1995 from the University of Bordeaux 1, where he is currently a professor at the Electronic Engineering School of Bordeaux (ENSEIRB). His main research activities, carried-out at the IXL laboratory, focus on mixed-signal circuit design and testing, digital and analogue signal processing, and programmable devices' applications. His interests include also digital design and its application in BIST structures for the characterization of embedded A/D converters, as well as, digital signal processing applied to nondestructive techniques based on time-frequency representation.

HADDADI, Djamel was born in 1971. He received the Engineer degree in physics and the Master degree in signal processing in 1996 from the Institut National Polytechnique de Grenoble, and the PhD degree in Electronics from the Université Bordeaux 1. He is a product responsible at STMicroelectronics since 2000. His main interests include ATE test and qualification of high performance analogue and mixed-signal ICs.

DURAND, Jacques obtained his PhD degree from the Université de Paris Sud — Orsay-IEF in 1984. Since then he worked at THOMSON-CSF (Orsay) as an expert on analogue signal processing and IC measurement and testing. His main interests concerned high-speed ADC test and characterization, with emphasis on noise and jitter analysis. Along his career he participated as an examinator in the evaluation of several PhD thesis, and published various conference and journal papers on ADC roadmap and characterisation. He holds a patent on a phase noise measurement set-up.

MACHADO da SILVA, José received the Licenciatura and PhD, both in Electrical and Computer Engineering from the Faculdade de Engenharia da Universidade do Porto (FEUP), Portugal, in 1984 and 1998, respectively. He joined FEUP in 1984 as a lecturer, and INESC-Porto in 1991 as a senior researcher, where he has collaborated on and supervised national and European projects on analogue and mixed-signal testing. He is currently an Assistant Professor at FEUP and a project leader at Instituto de Engenharia de Sistemas e de Computadores (INESC-Porto), with teaching and research responsibilities on design and testing of electronic circuits. His research interests include analogue and mixed-signal design for testability, new testing methodologies, analogue and digital signal processing, and VLSI design.

MARCHEGAY, Philippe was born in France in 1942. He received his PhD in 1966 and the "Doctorat en Sciences Physiques" degree in 1979 from the University of Bordeaux (France) about the topic of the study of metastability of synchronizer circuits and coherence faults of random access sequential networks. At present his research interests concern functional testing of A/D converters and their design. He is professor at the graduate engineering school, Ecole Nationale Supérieure d'Electronique Informatique et Radiocommunications de Bordeaux (ENSEIRB) of which he ensured the direction between 1999 and 2004.

MAZZOLENI, Sara was born in Milano, Italy, in 1967. She received the degree in Solid State Physics at the University of Milano in 1991. From 1990 until 1992 she worked as a researcher studying the non-linear behaviour of solid state lasers. From 1992 until 2001, she worked in Italtel SpA as test engineer of analogue and mixed-signal electronic components for telecommunication equipments. Now she is Supplier Quality Manager in Italtel Spa.

MENDONÇA, Hélio Sousa was born in Porto, Portugal, in 1968. He received the Licenciatura and PhD Degrees, both in Electrical and Computer Engineering from the Faculdade de Engenharia da Universidade do Porto (FEUP), Portugal, in 1991 and 2004 respectively. He joined FEUP in 1994, where he is now an Assistant Professor. He is also a Senior Researcher at Instituto de Engenharia de Sistemas e de Computadores (INESC-Porto) since 1995. His main research interests are digital signal processing and dynamic testing of ADCs.

MORANDI, Carlo graduated in Electronic Engineering at the University of Bologna in 1971. He worked at the University of Bologna as a research assistant, then as an associate professor of Electronic Instrumentation. Full professor of Applied Electronics at the University of Ancona since 1986, in 1988 he moved to the Faculty of Engineering of the University of Parma. His scientific interest is focused on the design and testing of mixed-signal integrated circuits and on the development of dedicated electronic instrumentation. He coordinated several national and international research projects, among them the "Standards, Measurements and Testing" project DYNAD of the European Commission concerning the definition of standard test procedures for the dynamic characterization of A/D converters, which originated the present book. He is author or co-author of over 100 scientific publications on international journals or proceedings of international conferences.

ROY, Pierre-Yves received the Engineer Diploma of the Ecole Nationale Superieure De Telecommunication de Bretagne ENSTB in 1995. He started his carreer working for Thomson-CSF (Thales now); first for Thomson-CSF Airsys (Thales Air Defence Systems) as a radar receiver designer, and then for Thomson-CSF Technologies and Methods (Thales research and Technology) as an expert in data conversion. When he was in Thales, his main areas of interest concerned high dynamic signal receivers and the functional testing of ADCs. In 2000, he joined EADS Telecom to manage the design of the architecture (and of the associated components) of their 3G secured radiocommunication terminals. He is now terminal architect for EADS Telecom.

Introduction

José Machado da Silva

ADCs are, eventually, the most pervasive analogue blocks in electronic systems. With the advent of powerful digital signal processing and digital communication techniques, ADCs are fast becoming critical components for system's performance and flexibility. Knowing accurately all the parameters that characterise their dynamic behaviour is crucial, on one hand to select the most adequate ADC architectures and characteristics for each end application, and on the other hand, to understand how they affect performance bottlenecks in the signal processing chain.

At present, most of the signal processing performed in electronic systems is becoming digital, and the role of the ADCs placed at the borders of the digital domain acquires a particular relevance, since the signal degradation introduced by these components cannot normally be recovered by subsequent processing. Both the markets of stand-alone ADCs and of ADC macrocells to be embedded in complex systems-on-chip, benefit from the availability of performance parameters accurately describing their expected behaviour, and of clearly specified test methods to be used for their measurement.

When the project DYNAD started, the standardization of ADC test procedures was not so well developed. Two standards existed, in particular, at that time — the IEC 60748 and the IEEE Std 1057. The former covers only quasi-static operation, while the second deals with dynamic testing but, being addressed at digital waveform recorders requires some adaptations to cover ADCs. A first aim of DYNAD project was then, to contribute to the improvement of the European rules concerning test methods for ADCs, by proposing an integration within IEC 60748 addressing the parameters specifying the dynamic behaviour of ADCs, measurement conditions, and data processing algorithms. By the end of year 2000 a working group from the IEEE Instrumentation and Measurement Society Technical Committee (TC-10) completed the IEEE 1241 Standard for Analog to Digital Converters. This standard, as well as

D. Dallet and J. Machado da Silva, (eds.), Dynamic Characterisation of
Analogue-to-Digital Converters, xv–xvi.
© 2005 *Springer. Printed in the Netherlands.*

contributions from the DYNAD project, are now being incorporated into an IEC standard on dynamic testing of ADCs. Other initiatives have been carried-out concerning standardization of ADC testing methods. One can also mention EUPAS (EUropean Project for ADC-based devices Standardization), and the IMEKO Technical Committee 4 (A/D and D/A Metrology WorkGroup).

The main objective of the DYNAD project was the study and evaluation of ADC testing methods based on the use of sinewave test stimulus. A second aim was to investigate and propose new test methods to circumvent the limits of the measurement instrumentation, which is strongly challenged by today's high resolution, high speed converters. Techniques for the measurement of parameters required by specific applications (e.g. audio hi-fi) and for the de-bugging of new converter designs were also investigated. Dissemination of the knowledge gathered during the activity was the third objective.

That work is now compiled in this book, which is structured in two main parts. Part one comprises chapters one to six. The first one provides an overview of the most important ADCs' architectures and respective fields of application. An introduction to the most relevant nomenclature and definitions of terms is also presented. Chapter two describes the generic architecture of an ADC test setup, and guidelines and best practice procedures are proposed in order to guarantee reliable test results. Chapters 3, 4, and 5 are devoted to the description of dynamic test techniques using sinewaves, respectively, sinewave fitting (time domain data analysis), discrete Fourier transform (frequency domain analysis), and code histogram test (statistical domain analysis). These techniques are thoroughly described, as well as the fundamental mathematical background behind the equations to be used to obtain ADCs' characterization parameters provided in each case. A comparison among these three methods is presented in chapter 6. The objective is not to find the best or the worst methods, but mainly to compare how they behave when test conditions are not ideal and to identify their requirements in terms of test time and volume of data. Examples of ATE implementation are also included.

The second part comprises chapters 7 to 10, which provide additional information to test for other relevant parameters, such as jitter, differential gain and phase, step and transient response, and hysteresis.

GLOSSARY

ε error, used for total error and error band

$\varepsilon_G(f)$ gain flatness error at frequency f

$\varepsilon[\mathbf{k}]$ difference between T[k] and the ideal T[k] computed from G and V_{os}

$\varepsilon_m(f)$ aliasing and first differencing magnitude errors

$\varepsilon_\theta(f)$ aliasing and first differencing phase errors

ε_q quantisation error

ε_{rms} root-mean-square value of ε

θ phase, expressed as radians

$\eta[\mathbf{n}]$ a record of noise data

η_f noise floor

π constant, ratio of the circumference to the diameter of a circle

ρ reflection coefficient

σ standard deviation; sometimes used as noise *rms* amplitude, which is the standard deviation of the random component of a signal

σ_σ standard deviation of the standard deviation (for example, standard deviation of the noise amplitude)

σ_j jitter

σ_t aperture uncertainty

σ^2 variance; sometimes used to describe random noise power

τ sampling period, the inverse of f_s

ω angular frequency, expressed in radians per second

ω_i angular input frequency in radians/second

$\delta\mathbf{t}_{eq}$ sampling time error of equivalent time sampling

$\delta\mathbf{t}_{fi}$ input frequency inaccuracy

A   sinusoidal amplitude

B   test tolerance in fractions of the nominal least significant bit (Q). Also used as an amplitude

BW   frequency bandwidth

c   general purpose constant

C   offset

d[n]   dither component of output sample y[n]

dest[n]   estimate of the dither component d[n]

D   general purpose integer

DFT   Discrete Fourier Transform

DG   differential gain

DNL[K]   differential non-linearity of code k

DNL   maximum differential non-linearity over all k

DP   differential phase

DR   dynamic range

ENBW   equivalent noise bandwidth

f   frequency, Hz

f(n)   sinewave component of output sample y[n]

$\mathbf{f}_{co}$ upper frequency for which the amplitude response is -3 dB

$\mathbf{f}_d$ sampling frequency of a record after decimation by some integer D

$\mathbf{f}_{eq}$ equivalent sampling rate

$\mathbf{f}_h$ frequency of an harmonic of the input frequency

$\mathbf{f}_i$ actual input frequency or approximate desired input frequency

$\mathbf{f}_{imf}$ frequency of intermodulation distortion products

$\mathbf{f}_{opt}$ optimum input frequency for testing

$\mathbf{f}_m$ frequency of the mth component of a magnitude spectrum

$\mathbf{f}_r$ input signal reference frequency or input signal repetition rate

$\mathbf{f}_s$ sampling frequency

$\mathbf{f}_{sp}$ frequency of a persistent spurious tone

FR   frequency response

G   static gain of the ADC under test

G(f)   dynamic gain of the ADC under test as a function of frequency

h   order of harmonic frequency

H   average number of histogram samples received in two code bins sharing
the same transition level

H[f]   frequency response of the ADC under test

H[i]   number of histogram samples in bin i

Hc[j]   number in the jth bin of the cumulative histogram of samples

$\mathbf{H[f_k]}$ DFT of h(n)

h(n)   discrete time impulse response of a system

i   general purpose index

I   general purpose factor

IMD   intermodulation distortion

INL   integral non-linearity

INL[k]   integral non-linearity at output code k

J   number of cycles in a record

k   code bin

L   general purpose integer

L(f)   phase noise spectral power density

mse   mean square error

M   number of sequential samples in a record

M$^+$(x), M$^-$(x)   number of measurements of the output value at the input value x for increasing and decreasing inputs respectively

Md   number of samples in a record after decimation

MD   number of samples in one period of pseudo-random dither

n   sample index within a record

N   number of bits

N$_{ef}$ number of effective bits

NDR   noise distortion ratio

NPR   noise power ratio

p   probability

PG   processing gain

Q   ideal code bin width, expressed in input units

r   general purpose integer

R   minimum number of records required

S   set of samples collected over more than one record, also used as an error parameter or as total number of samples used in a histogram

SFDR   spurious free dynamic range

SINAD   signal to noise and distortion ratio

SNR   signal to noise ratio

$S_x(f)$ spectral power density of quantity x

t$_{eq}$ average equivalent time sampling period

t$_f$ top to base transition time; falltime

t$_r$ base to top transition time; risetime

t$_n$ discrete sample times

t$_{wc}$ the center point of the aperture time associated with an output sample

T[k]   code transition level between codes k-1 and k

T$_{nom}$[k]   nominal code transition level between codes k-1 and k

TĤD   estimate of total harmonic distortion

THD   total harmonic distortion

TSD   total spurious distortion

u   confidence level expressed as a fraction

V$_{cm}$ common mode signal

V$_{dm}$ differential mode signal

V$_{fs}$ full scale range

$\mathbf{V}_{fsn}$ nominal full scale range

$\mathbf{V}_{ifs}$ full scale input signal

$\mathbf{V}_{OS}$ ADC input offset, ideally $= 0$

$\mathbf{V}_{OD}$ input signal overdrive; the amount by which an input signal exceeds the ADC full scale range

$\mathbf{V}_{rir}$ reduced ADC input range

VSWR   voltage standing wave ratio

w   estimated code error rate

w'   worst-case code error rate

w[n]   window function coefficient (for a DFT)

W[k]   code bin width of code bin k

x   ADC input signal value; or number of errors detected

X   number of standard deviations of a Gaussian distribution

$\mathbf{X}_{avm}(\mathbf{f}_m)$ the averaged magnitude spectral component at discrete frequency f_m after a DFT

y[n]   the nth output data sample within a record

$\overline{y}[\mathbf{n}]$ average of y[n] over M samples

yn'   best fit points to a data record

Y[k]   the k-point DFT of the M-sample record y[n]

$\mathbf{Z}_O$ transmission line impedance

$\mathbf{Z}_t$ ADC input impedance

Zu/2   number of standard deviations that encompass 100(1-u) % of a Gaussian distribution about the center.

I

ADC CHARACTERISATION BASED ON SINEWAVE ANALYSIS

Chapter 1

ADC APPLICATIONS, ARCHITECTURES AND TERMINOLOGY

José Machado da Silva
Universidade do Porto, FEUP – INESC-Porto
Campus da FEUP, Rua Dr Roberto Frias
4200-465 Porto, Portugal
jms@fe.up.pt

Hélio Mendonça
Universidade do Porto, FEUP – INESC-Porto
Campus da FEUP, Rua Dr Roberto Frias
4200-465 Porto, Portugal
hsm@fe.up.pt

1. Introduction

An analogue-to-digital converter (A/D converter or ADC) takes a continuous analogue input signal, most often a voltage V_x, and converts it into an N-bit binary number, which can then be manipulated by a computer. This functionality leads ADCs to present a very important role in electronic systems for a wide range of applications. Traditional mixed-signal implementations comprehend high-performance analogue circuits combined with a few digital functions for control or interface — these are the so-called *register-controlled analogue* systems. With the development of semiconductors' technology and microprocessors, systems based in digital signal processing cores and provided with analogue inputs and outputs — the so-called *digitised analogue* systems — are becoming more and more the basis of the dominant architectures in electronic systems. The performance of the ADCs placed at these analogue/digital interfaces acquires a particular relevance, since the effects of the ADCs' behaviour on the signals being acquired, cannot easily be recovered by subsequent processing. Additionally, the performance of most recent systems for applications in fields such as telecommunications, test and measurement, or consumer

3

D. Dallet and J. Machado da Silva, (eds.), Dynamic Characterisation of
Analogue-to-Digital Converters, 3–45.
© 2005 *Springer. Printed in the Netherlands.*

electronics require ADCs with ever increasing capabilities in terms of resolution and speed.

To optimise a system's performance, designers using high-performance A/D converters must thoroughly understand their test techniques and limitations. The traditional ADCs' performance characterisation parameters obtained with static tests are not sufficient to fully characterise the device's behaviour. One such static test consists in placing the device under test within a DC servo-loop, in order to measure each code transition level individually [1]. This method is still quite useful to obtain the DC transfer characteristic of high resolution, low sampling-rate converters, and then to calculate parameters like integral and differential non-linearities, and monotonicity. However, instead of a steady DC level, the A/D converter is in general subject to a dynamically changing input signal. The best test approach would be to evaluate the ADC performance in the target system, but this is not always feasible. In this case the converter should be tested under conditions as close as possible to those required for the final application. Dynamically testing an ADC in the frequency range that covers the bandwidth of the application can provide that assurance, being the ADC performance given now in terms of dynamic or AC specifications.

Three test methods are commonly used to dynamically characterize an ADC — histogram analysis, spectrum analysis, and time-domain analysis (also known as sine wave curve fitting). These three methods operate essentially in the same manner, i.e., a sine wave stimulus is applied to the ADC and one or more records of data are taken from the ADC output response, which are then processed to extract relevant parameters. As these methods differ in the data processing algorithms, and consequently in the type of errors detected, they do not provide, at the first approach, the same characterisation parameters. A fourth method — the beat frequency method — is often used as well, which is able to provide a quick yet simple visual demonstration of the ADC dynamic behaviour. However, as it is a qualitative test, it is not usually used as a reference method to provide numerical characterisation data.

Concerning the selection of the input stimulus waveform, the ramp is the simplest one for testing an ADC's linearity. This signal presents however a drawback that comes from the fact of being hard to generate with the proper linearity. The input stimulus ramp should have at least two to four bits of resolution more than that of the ADC under test, a requirement which might be difficult to meet when testing high-speed or high-resolution converters. Alternatively, a sine wave can overcome these problems, although its purity is critical to the success of the tests. A spectrally pure sinusoid is easier to obtain than a sharp and perfectly linear triangular wave, by taking a lower resolution sine wave and filtering it to the required purity. Often, synthesized sources are necessary to provide the short-term and long-term stability required by the dynamic range of the ADC. Anyway, precisely controlling the absolute amplitude and offset of a sinusoid waveform is not trivial.

A brief introduction to the problem of dynamically testing A/D converters was given above. One of the aspects to be taken into account is the necessity of testing the converter as much as possible under the same conditions as those found in the final application. As this is not usually possible, a central question concerns the identification of the ADCs' performance characterisation parameters that best evaluate its ability to perform the desired function.

In fact, often the generation of ADC tests is driven by the application characteristics or by the specificities of the ADC architecture, in order to adapt specification parameters limits, the test setup, or even the test methods themselves. For example, for certain applications absolute converted digital code combinations may not be as important a characteristic as relative converter response to the adjacent code combinations. This is the case, for instance, in digital audio applications where two important ADC performance characteristics are the non-linearity causing signal distortion and the intermodulation of high frequency tones to the baseband. The high resolution converters used in these applications require clock jitter to be considered for it causes a signal dependent modulation and additional noise. It is also important to avoid group delay differences which influence focus and stability of the sound sources in a stereo image — humans can detect time-of-arrival differences of about 7 microseconds [87]. Performance at low levels and at higher frequencies is vital for good sound quality.

This chapter continues with an overview of different application fields for ADCs, and the respective most critical parameters. Section 3 presents the most commonly used ADCs' architectures and the respective current rates of number of bits and sampling frequencies. Sections 4 to 9 address the non-harmonised approach seen today concerning the terminology used in the analogue-to-digital conversion domain, present definitions for different terms, and introduce acronyms used along this book. The information given herein was obtained from an extensive bibliography, as well as, from manufacturers data sheets and application notes. It is considered to be, at the time of writing and at the best of our knowledge, an updated state of the art overview of the technology in this domain. However, as technology is continuously developing, it is likely that part of this information becomes out-of-date.

2.   ADCs' applications

This section presents an overview on ADCs' most significant application fields, and for each one identifies the parameters which are the most critical for the overall system's performance.

Table 1.1.   Critical ADCs' performance parameters per application.

Application	Critical Performance parameters	Performance issues
Audio	SINAD, Crosstalk	Power consumption Gain matching
Automatic control, Sensors and robotics	Monotonicity, short term setting, linearity, long-term stability, temperature offset	Transfer function
Data transmission	SFDR, BW, SINAD DR, INL, DNL	Thermal noise Phase non-linearity
Digital high-Speed Instrumentation	N_{ef}, BW, out-of-range recovery, Word error rate	SNR for better wide bandwith amplitude resolution, SFDR to minimize distortion, bit error rate thermal noise
Geophysical	THD, SINAD, DR Long-term stability	miliHz response
Hard disk driving	Conversion time/latency	
Medical	SFDR, BW, INL, DR, SNR	

Table 1.1 (continued)
Critical ADCs' performance parameters per application.

Application	Critical performance parameters	Performance issues
Military communications	SFDR, SINAD, THD, IMD, NPR, NDR	Linear dynamic range for detection of low-level signals in a strong interference environment
Electronic warfare	SFDR, SINAD, NDR	Sampling frequency
Mobile telecommunications and wireless communications	SINAD, NPR, SFDR, THD, SNR, IMD, NDR	Wide input bandwidth channel bank Bit error rate, word error rate Interchannel crosstalk, compression, power consumption
Monitoring, test equipment and instrumentation	N_{ef}, BW, out-of-range recovery, word error rate	SNR for better wide bandwidth, amplitude resolution, SFDR to minimize distortion, bit error rate
Radar and sonar	SINAD, SFDR, INL, BW Out-of-range recovery	SINAD for clutter cancellation and Doppler processing
Spectrum analysis	SINAD, SFDR	SINAD and SFDR for high linear dynamic range measurements
Speech and voice communications	SINAD, NPR	
Video and television	INL, DNL, FR, SNR, DG, DP, SFDR, word error rate, BW, THD, SINAD	Differential gain and phase errors Power consumption

Table 1.1 presents a summary of ADCs' performance characterisation parameters and other behavioural aspects which are critical for different application fields. Regarding the selection and definition of these fields, some of them are somewhat similar in their essence but, due to significantly different performance requirements imposed on the ADCs used in the different applications, they were considered as distinct. This is the case, for instance, of audio and speech and voice communications applications, in which the later places less strict requirements (lower resolution and conversion rate) on the ADCs to be used. Similarly, ADCs for hard-disk drives are described as belonging to an application field of their own due to the exceptionally high conversion rates required (much higher than those in other control applications such as in automatic control and robotics). The same option was made for ADCs for geophysical applications, in which the required resolution is higher but the conversion rate lower. Also on this line, digital high-speed instrumentation, which comprises mainly instruments such as digital oscilloscopes, was considered a different application field from other test and monitoring equipment due to the higher performance requirements. Spectrum analysis was also considered a field apart from generic test and monitoring equipment, because it has precise requirements regarding the accurate conversion of different frequency components and the ability to detect even low power signals, spectral noise density and linearity being two key specifications in these applications.

In high-speed applications, such as video and wireless communications, whether they require undersampling or oversampling, the ADC must deliver high levels of dynamic performance. Since it generally is the front-end of the application, the overall system specifications will depend on the ADC's dynamic performance, characterized by parameters such as SFDR, THD and SNR [20]. SFDR is important in many applications because noise (including thermal noise) and harmonics restrict the dynamic range. In an IF (intermediate frequency) bandpass converter, for example, spurs may be interpreted as adjacent-channel information, giving rise to inter-channel mixing because different channels reside relatively close to each other.

The same can happen in military applications, such as signal intelligence and communications intelligence, where distortion within the ADCs, can lead to false readings [49]. High SFDR and low THD help to minimise the ADC's contribution to the overall distortion. In yet other applications, signals of interest may not be distinguishable from harmonics and spurious signals. In echo-cancelling modems (where the modem transmits and receives at the same time) the ADC must have sufficient dynamic range to capture the strong echo of the "send" signal and the weak "receive" signal (40-50 dB weaker than the "send" echo in long lines) without clipping the echo and without loosing the weak "receive" signal in the converter's quantisation noise. Echo clipping would cause distortion and ruin the signal because a synthesized copy of it is subtracted from the complete captured signal. This may require up to 1 MHz of band-width and up to 90 dB of dynamic range [94]. In high speed modems, where data is coded by phase-shift keying, phase non-linearity (or delay distortion) is also an issue.

In audio applications, particularly in stereo systems, phase non-linearity gives place to time-delay distortion which, for an input signal consisting of two or more frequencies, renders distortion caused by the different arrival times of these frequency components at the output of the ADC.

Radar applications demand the best possible SFDR and SINAD in order to prevent weak signals from being masked by harmonics or spurious signals. The harmonics responsible for poor radar SINAD performance (typically the second or third ones) arise due to poor INL at some particular frequency. Thus, bandwidth and slew rate can be key factors. Sonar is another application which also requires low noise and good SFDR. Geophysical sensor equipment re-quires high dynamic range. The signals which result from the execution of acoustic geophysical tests easily span dynamic ranges of over 120 dB [67].

Spread-spectrum techniques are used in both military (subject to very strong interference) and wireless communications where there are multiple users in the same frequency in overlapping networks. ADCs for these applications are required to present good SINAD, IMD, SFDR, and NPR characteristics with performance issues being low IMD for quantisation of small signals in a strong interference environment, low SFDR for spatial filtering, and high NPR for low interchannel crosstalk. It is important to successfully reject adjacent channels. Medical imaging applications, such as ultrasound systems, require ADCs with good dynamic performance at high sampling rates. Ultrasound systems' ADCs need wide dynamic range to prevent noise from masking subtle abnormalities in a diagnostic image. In these applications ADCs' useful bandwidth is deter-mined by the amount of SFDR needed within the system.

Imaging and video applications need excellent linearity (DNL and INL). In fact, DNL is the most important specification parameter for the ADCs used to capture CCD (charge coupled devices) sensor signals. DNL affects intensity fidelity and causes improper gradation of the image scale with local imper-fections. INL corrupts the entire image with a gradual non-linearity which in case of colour images may result in colour artifacts that may cause deceptive results [127]. They also require good low-noise (SNR and SINAD) perfor-mance. Some tasks require converters with a fast slew rate sample and hold amplifier to handle a full-scale step response from pixel to pixel. SNR over a wide bandwidth is important in the case of applications which may spread this change over several pixels [66]. SNR is also critical in medical imaging [49] .

Low power consumption is an increasingly important specification in every application fields, but it finds particular importance in those where portability is required, such as in camcorders (video), cellular phones (personal communications), walkmans (audio) and portable instrumentation (hand-held measurement and monitoring equipment), where it is important to achieve ever greater autonomy with existing batteries.

3.   ADCs' architectures

Table 1.2 presents the most commonly used ADCs' architectures and their respective resolution and conversion (output data) rates for the same application fields depicted in table 1.1. Converters based on $\Sigma\Delta$ modulation followed by decimation are the most popular for audio applications. Due to their conversion mechanism, the $\Sigma\Delta$ converter is inherently monotonic and presents very low DNL. Some of the additional advantages are modest circuit accuracy and component matching and trimming requirements, which make them suitable for implementation using imprecise CMOS technologies. Also, the need for external anti-aliasing filtering is reduced by the (digital) decimation process. The requirement for sampling rates up to 48 kHz, 96 kHz, or even 192 kHz for more demanding applications, such as professional audio, are well within the possibilities of even very high resolution $\Sigma\Delta$ converters. Some other high resolution low sampling rate applications, such as geophysical, make also good use of the advantages of $\Sigma\Delta$ converters. In this latter case, however, the order of the $\Sigma\Delta$ modulator must be increased in order to allow for lower oversampling ratios. This trade-off between oversampling ratio and modulator order is also found in $\Sigma\Delta$ converters for other applications that require relatively high sampling rates (although with lower resolution), such as, sensors, instrumentation, GSM (Global System for Mobile communication), and data transmission. Fourth order, 13-bit, 270 kS/s $\Sigma\Delta$ converters are used in GSM because the high sampling rate offers the necessary bandwidth to pass the adjacent channels and the blocking levels without aliasing them in the band of interest, and the decimator filter besides providing downsampling also performs channel filtering . The worst case blocking specs of the GSM standard require a conversion linearity of 14–16 bits to avoid a weak received signal being lost due to distortion artifacts [125]. Also, high dynamic range (≈ 80 dB) and SNR ranging from 86 to 98 dB are fundamental to allow demodulating low level signals immersed in strong adjacent channel interfering signals [76].

Anyway, $\Sigma\Delta$ have somewhat high power consumption levels when compared with converters of similar resolution and conversion rate using other architectures (pipelined, for instance). Cost and power consumption issues require the use of such specific architectures to achieve high resolution and conversion rate at minimum power consumption, especially in the case of portable, hand-held devices. Equipment of this sort can mainly be found in applications such as mobile and wireless communications, instrumentation, and video. Latency and larger size are two other disadvantages to be taken into account.

Table 1.2. Characteristics of ADCs per application.

Application	Converter architecture	Resolution (bits)	Conversion rate
Audio	ΣΔ	14-18 for consumer audio	48-50 kS/s
	ΣΔ, 4th-7th order ΣΔ	18-24 for professional equipment	48-96 kS/s
	SAR	10-16	85 to 500 kS/s
Automatic Control,	ΣΔ	24	780 S/s
Sensors and	SAR	8-18	20-2000 kS/s
Robotics	Integrating	18-20	100-2000 S/s
	ΣΔ	16	192 kS/s
	Half-flash	8	200-400 kS/s, 1MS/s
	(for high-speed servo control)		
Data transmission	ΣΔ and high-order	12-16 for cancelling modems	8 kS/s for modems
(cable)	(4th) ΣΔ for ISDN[a] and ADSL[b]	13-16 for ISDN	80-160 kS/s for ISDN
	Pipeline	12 for ADSL	2.2 MS/s for ADSL
		12 for VDSL[c]	40 MS/s for VDSL
(optical fiber)	Half-flash	8-12	400 kS/s-1.5 MS/s
Digital high-speed	Flash, interleaved flash	8	150 MS/s-1 GS/s
instrumentation		10-12	10-40 MS/s
	Flash-SAR hybrid	12-16	85-166 kS/s
	SAR	8-18	20-2000 kS/s
	Pipeline	8-14	1-80 MS/s

[a] Integrated services digital network
[b] Asymmetrical digital subscriber loop
[c] Very high data rate digital subscriber loop

Table 1.2 (continued)
Characteristics of ADCs per application.

Application	Converter Architecture	Resolution(bits)	Conversion rate
Hard disk driving	Half-flash	10	320 kS/s
	Pipeline	8-12	800 kS/s-1.5 MS/s
	SAR	8	100 kS/s
	Flash	6	30-140 MS/s
Medical	Interleaved flash (for CAT[a])	8	150-750 MS/s
	Flash	8	1 GS/s
	SAR	8-18	20-2000 kS/s
	ΣΔ	14-16	200-10000 kS/s
	Pipeline	8-14	1-80 MS/s
Military communications	Flash, interleaved flash	8	150-750 MS/s
	Subranging, pipeline,	10	1.5 GS/s
Electronic warfare	Folding, and interpolating	12-14	50-100 MS/s
		8	3 GS/s
Mobile telecommunications and wireless communications	SAR for GSM	8 for GSM	270 kS/s for GSM
	ΣΔ and high-order (4th)		
	ΣΔ for GSM	13 for GSM	
	Half-flash	8-10	320-500 kS/s, 1 MS/s
	Pipeline	8-14	1-80 MS/s
	Flash	6 for satellite	40-80 MS/s
	Flash and interleaved flash	8	150 MS/s-1 GS/s for RF
	Pipeline	12-14	48-65 MS/s for IF

[a]Computer axial tomography

Table 1.2 (continued)
Characteristics of ADCs per application.

Application	Converter architecture	Resolution(bits)	Conversion rate
Geophysical	ΣΔ	16-24	1-32 kS/s
Monitoring, test equipment and instrumentation	ΣΔ Half-flash SAR ΣΔ	22-24 8 8-18 14-16	down to < 1 kS/s 400 kS/s 20-2000 kS/s 200-10000 kS/s
Radar and sonar	Flash and interleaved flash Pipeline Subranging, pipeline, Folding and interpolating for radar ΣΔ for sonar	8 12 12-14 for radar 16-18 for sonar	150 Ms/s-1 GS/s 10-30 MS/s 50-100 MS/s for radar 200 kS/s for sonar
Spectrum analysis	Pipeline SAR	10-12 12-14	10-40 MS/s 300-500 kS/s
Speech and voice communications	SAR ΣΔ	8-18 11-14	20-2000 kS/s 8 kS/s
Video and Television	Half-flash for professional video Pipeline	8-12 8-12	10-40 MS/s 30-50 MS/s

In the first GSM generation 8 bit SAR (successive approximations) converters were used. Comparing to $\Sigma\Delta$ these converters present the advantage of joining smaller size, lower power, and lower latency, but on the other hand require stringent trimming to achieve good accuracy and more tight front-end anti-aliasing filtering. Automatic control is a field where their fast sampling with no latency and good resolution match well together.

Data communication is another application that demands frequently low power. Architectures for data transmission, with the increase in resolution and conversion rate requirements, have migrated from SAR and flash to $\Sigma\Delta$, and pipeline. $\Sigma\Delta$ converters are used for instance in ISDN, with 3rd order architectures achieving 16 MHz sampling frequencies. In ADSL, due to the high Nyquist rates needed (2.2 MHz conversion rate), high order (4th) $\Sigma\Delta$ converters are currently being used. In order to reduce power consumption pipelined converters with 8.8 MHz sampling frequency (2.2 MHz conversion rate) will be used in the future, the trade-off being stricter production tolerances. $\Sigma\Delta$ converters are not an option for VDSL due to the high speed requirements. Full flash, subranging, folding and interpolating, and pipelined ADCs are better alternatives, the pipelined being the best option as it is possible to attain 10-12 bits resolution at 10-100 MS/s conversion rates. Pipelined architectures are also ideal for A/D converters requiring both high speed and high resolution [98], thus finding preferred use in applications such as video, data communications (for ADSL and VDSL) and medical imaging. Folding and interpolating converters, being capable of higher sampling rates but lower resolutions, are also becoming an alternative in these areas. Folding and interpolating 8-bit 3 GS/s, and pipelined 10-bit 1.5 GS/s state-of-the-art ADCs for defense applications have been reported [107].

In wireless communications the current trend is towards moving the signal quantisation procedure into the RF (radio frequency) stage and performing IF filtering in a digital form (the concept of "software radio") [86, 144]. This obviously creates the need for extremely high conversion rate converters with high SFDR to avoid interchannel mixing because the quantiser precedes the IF filter. This is an example of a perfect application field for parallel (flash) or time-interleaved pipelined converters, in which several pipeline converters sample the same signal in intermediate time instants, thus achieving a higher sample rate than a single converter with the wide dynamic range characteristics of slow converters.

Flash converters, from a power dissipation perspective, are acceptable only at low resolution (up to 8 bits) levels where the number of comparators is relatively small and their offset is non-critical — every extra bit requires doubling the comparators in number and accuracy. At resolutions in the 8-12 bits range, the only practical options for low power dissipation are multistep flash and pipeline configurations. Multistep flash has been successfully used in low-power applications at the 10-bit level. Pipeline are also attractive and have the potential advantages of inherent single-path sampling of the signal, giving good high-frequency effective bit performance, and the capability of using non-critical purely dynamic comparators because of the amplification of the signal in the pipeline coupled with the use of digital correction. In video, for digitisation of the CCD sensor array information, and due to the amount and update rate of the same, a high-speed converter is necessary — flash (in the past) and pipeline being the preferred converter architectures.

4.   Terminology

In the ADCs' domain one can often find different terms meaning the same thing, or to find different definitions for the same terms. Currently, different terms and acronyms are likely to be found among data sheets from different ADC manufacturers — e.g., while some use DNL for Differential Non-Linearity, others use E_D [126, 153]. To designate the effective number of bits acronyms like ENB, $NOEB$, $ENOB$, E, b_{ef} are likely to be found. In this case we suggest to use N_{ef} as N is already used to identify the number of bits. In order to have a single symbol to identify common parameters, η is used to designate parameters or quantities related to noise, ε is used for those related to errors, and σ for those concerning deviation. For example, η_f, ε_G, and σ_j, would identify respectively, noise floor, gain flatness error, and jitter.

Also, commonly the parameters specified in ADCs' data sheets are different for different manufacturers, being that dynamic characterisation parameters are not always specified. It is likely that only one of signal to noise ratio (SNR or S/N) and signal to noise and distortion ratio (SINAD) is specified. Frequently, the parameter total harmonic distortion plus noise (THD+N) is used instead of SINAD. Many manufacturers define the same specification differently, or use different methods to evaluate the parameters. For example, the range (number) of frequencies which are considered as harmonics (excluding DC) varies — some use the first ten harmonics [6], others only the second through the sixth harmonics [151], but specific applications may require considering all harmonics in the frequency range of interest.

This difference among terms and definitions used by different parties can also be found between standards. Table 1.3 presents a résumé of the terms described in standards IEEE-STD-1241 [6] and IEC 60748-4 [3]. At a first glance one can see that the IEC standard addresses mainly static performance characterisation terminology.

Looking at the definitions used in these two documents one can see that, in general, terms addressed by both are defined in a different manner. See, for example, the definition given for analogue-to-digital converter:

IEEE *A device that converts a continuous time signal into a discrete-time discrete-amplitude signal.*

Table 1.3.   Comparison between IEEE 1241 and IEC 60748-4 terminologies.

IEEE 1241	*IEC 60748-4*
	absolute accuracy error, total error
AC-coupled ADC	
alternation band	
	conversion code
analog-to-digital converter (ADC)	analog-to-digital converter (ADC)
	linear ADC, non-linear ADC
aperture delay	
aperture uncertainty	
common-mode rejection ratio	
code bin k	step
code bin width W[k]	step width
ideal code bin width (Q)	
code transition level	
code transition level T[k]	transition value
coherent sampling	
common-mode out-of-range	
common-mode out-of-range recovery time	
common-mode signal	
	conversion time
crosstalk	
differential input impedance to ground	
differential non-linearity	differential linearity error
differential signal	
epoch	
equivalent-time sampling	
full-scale range	full-scale ranges
	full-scale, zero-scale
	nominal full-scale value
full-scale signal	
full-width-at-half-max (FWHM)	
	full-scale error
	zero-scale error

Table 1.3 (continued)
Comparison between IEEE 1241 and IEC 60748-4 terminologies.

IEEE 1241	IEC 60748-4
gain and offset	gain, offset point, gain point
(A) (independently based)	
(B) (terminal-based)	
	offset error, gain error
harmonic distortion	
hysteresis	
incoherent sampling	
input impedance	
integral non-linearity	(end-points) linearity error
	best-straight-line linearity error
kth code transition level T[k]	
large signal	
least-significant bit (LSB)	LSB
long-term settling error	
maximum common-mode signal level	
maximum operating common-mode signal	
maximum safe input signal level	
	missing code
monotonic ADC	monotonicity
noise (total)	noise
normal mode signal	
equal to the differential signal	
overshoot	
out-of-range input	
passband	
phase non-linearity	
pipeline delay	
precursor	
probability density function	
quantization	
quantization error/noise	inherent quantization error
random noise	

Table 1.3 (continued)
Comparison between IEEE 1241 and IEC 60748-4 terminologies.

IEEE 1241	*IEC 60748-4*
record of data	
relatively prime	
residuals	
	resolution
	roll-over error
root-mean-square (rms)	
root-sum-square (rss)	
sampling	
	conversion rate
settling time	
short-term settling time	
signal-to-noise and distortion ratio (SINAD)	
signal to full scale ratio (SFSR)	
signal to non-harmonic ratio (SNHR)	
single-ended ADC	
slew limit	
spurious components	
spurious-free dynamic range (SFDR)	
step (or pulse) baseline	
step response	
step (or pulse) topline	
synchronous and asynchronous sampling	
timing jitter	
timing phase	
total harmonic distortion (THD)	
total spurious distortion	
transfer curve	
transition duration of a step response	
voltage standing wave ratio	
window	
word error rate	

IEC *A converter that uniquely represents all analogue input values within a specified total input range by a limited number of digital output codes, each of which exclusively represents a fractional part of the total analogue input range.*

Note - This quantization procedure introduces inherent errors of $1/2$ LSB (LSB - least significant bit) in the representation since, within this fractional range, only one (input) analogue value can be represented free of error by a single digital output code.

The definition given in the IEC standard is more precise on specifying a limited range for the input signal and for the number of digital output codes. Anyway, this definition is at a certain extent not complete as it addresses only the case of deterministic conversion laws. Other examples can be found in the definitions of *full-scale range*, *least-significant bit*, *quantization error*, and *offset* and *gain error*.

The following sections of this chapter will present the terminology and definitions that will be used along this book. These were adopted taking into consideration the section *Definitions* of the IEEE 1241 standard [6] and the section *General terms* of the IEC 60748-4 standard [3]. However, new terminology and symbols were adopted whenever it was found necessary. For example, the following terms are not found in the definitions sections of either the IEEE or the IEC documents:

- Effective number of bits

- Intermodulation distortion

- Decimation

- Histogram

- Average output code

- Input units

- Overdrive

- Signal to full-scale ratio

- Noise floor

- Envelope delay distortion

- Nominal code transition level

- Processing gain

- Reduced input range

- Dither

- Jitter

- Noise power ratio

- Gain flatness error

5.   Quantisation and A/D conversion

5.1   Analogue-to-digital conversion law

The mapping of all analogue values of the input quantity $x(t)$ falling within a specified full scale range V_{fs}, to a finite number k of digital output codes belonging to the set of possible output codes. The input quantity is usually a continuous-amplitude, continuous-time signal $x(t)$, and the mapping must be evaluated at selected time instants, specified by a convert signal.

This term may be considered similar to *Transfer characteristic*, however this one represents the particular case of the graphical representation of a deterministic conversion law.

See also: Transfer characteristic (5.5), Full-scale range (9.2).

5.1.1   A/D deterministic conversion law.   A particular form of probabilistic conversion law, where the conditional probability $p(k, x)$ is 1 (certainty) if x belongs to one or more intervals of the x axis, and 0 when outside of such interval(s). In other words, the x axis is partitioned into a set of adjacent intervals, each of which is univocally associated to a specific output code k. The interval (or union of intervals) associated to output code k is called **code bin k** or **quantisation cell k** (see 6.3).

This situation corresponds to the one outlined in figure 1.2. The result of the conversion is k if and only if x belongs to code bin k. It is assumed that the union of the possible code bins covers the entire input range V_{fs} (some result will surely be generated). Only in abnormal cases (non-monotonic conversion curves) can the same output code be associated with non adjacent intervals. Apart from such cases, it can be assumed without loss of generality that the codes k associated with intervals corresponding to increasing values of x represent increasing integer numbers. A common way of specifying a deterministic conversion law is by means of the transfer characteristic.

See also: Transfer characteristic (5.5), Monotonic ADC (6.10).

5.1.2   A/D probabilistic conversion law.   An A/D conversion law specified by the conditional probabilities $p(k/x)$ of output code k once the value of the input is x[1].

Conditional probabilities $p(k/x)$ are commonly referred to as **channel profiles**, with a term deriving from the nuclear instrumentation environment. The

---

[1]Note that this definition entails a memory-less model of the A/D conversion process. It is admitted that once the value of $x(t^*)$ at the sampling time t^* is known, it is possible to predict the probability of each output code. This is not true particularly at the highest sampling speeds: a real-world A/D converter is a complex, non-linear dynamic system that does not match the memory-less model. In order to predict the result of the conversion, it is also necessary to take the internal state of the A/D conversion circuit into account, which in turn reflects the past history of $x(t)$ and of the convert signal.

shape and position along the x axis of each of the conditional probability functions $p(k/x)$ is referred to the value of full scale V_{fs}. If all the $p(k/x)$ functions have the same shape and differ only for a translation along the x axis, the ADC is said to be **linear** (see 6.11).

A few noteworthy shapes of $p(k/x)$ are shown in figures 1.1, 1.2, and 1.3. The symmetric triangular shape of figure 1.1 is typical of time interval mea-

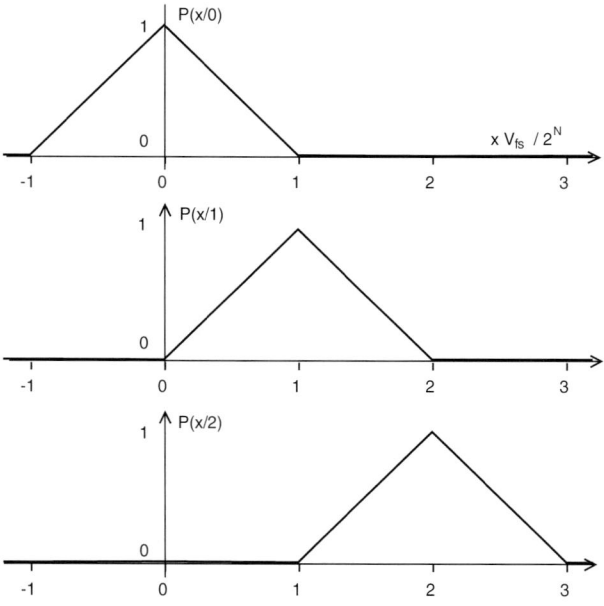

Figure 1.1. Triangular channel profiles

surements performed by counting the number of positive (or negative) edges of a clock signal with period T_{clock} falling within the interval. It has the property that, in the case of repeated measurements, the expected value of the measurement result is exactly the unknown duration of the time interval, so that the arithmetic average of several measurement results is an unbiased estimator of the interval duration.

The rectangular shape of figure 1.2 is typical of deterministic conversion laws (see 5.1.1) where the unknown input x is compared to a certain number of code transition levels which partition the x axis in disjoint intervals. In real world, due to the presence of unavoidable noise sources in the circuit, the electrical quantities representing the code transition levels are affected by noise, so that when the input is close to the nominal position of a transition level, there is a finite probability of misclassification, leading to the adjacent code. Thus, a more realistic representation of $p(k/x)$ for such a converter would be of the type of figure 1.3, which intuitively supports the choice of defining code transition

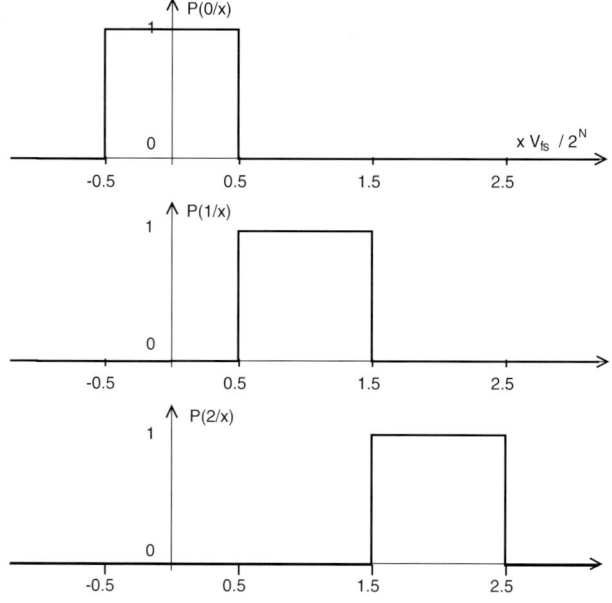

Figure 1.2.   Rectangular channel profiles

level $T[k]$ as the value of the converter input which causes half of the digital output codes to be greater than or equal to k, and half less than k. For a more correct and formal definition, it is necessary to refer to the theory developed in [43].

See also: Code transition level (6.5), Code error rate (7.1).

5.2   Analogue-to-digital converter (ADC)

A device that converts a continuous amplitude, continuous- or discrete- time signal $x(t)$, falling within a specified full scale range V_{fs}, into a discrete-amplitude, discrete-time signal, according to an assigned A/D conversion law that represents all analogue input values by a limited number of digital output codes, each of which representing a fractional part of the total analogue input range (see figures 1.1 to 1.3, for example). A typical ADC includes:

- an analogue (differential or nondifferential) input port, where the physical quantity representing the input signal x is applied

- a reference port, to which an external or built-in reference source V_{ref} is connected, which provides a physical quantity to be compared, after suitable scaling, with the input

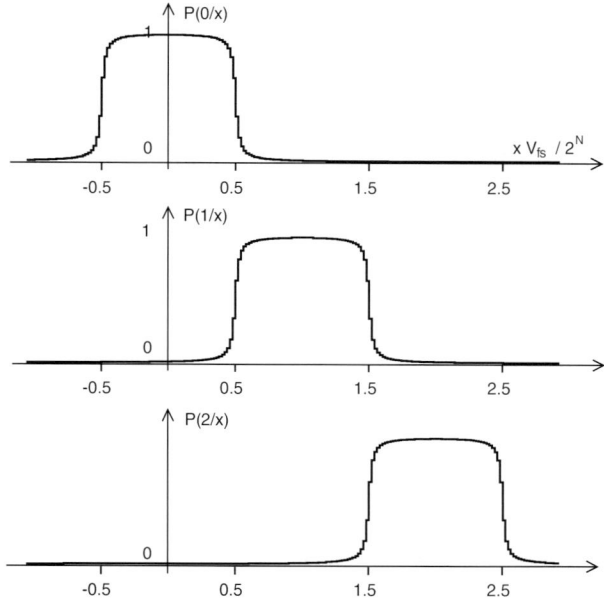

Figure 1.3. Smoothed rectangular channel profiles

- a convert command port, where the signal defining the start of the conversion process is applied

- an end of conversion port, that signals the end of the conversion process.

- an output digital port, where the digital code resulting from the conversion may be collected at the end of the conversion.

Central to the concept of ADC is the assumption that, despite the inherent amplitude and time errors introduced by the quantisation process, the information obtained at its output is equivalent to the one fed into the input. This, however, imposes limits. An ideal ADC is essentially an impulse sampler. This impulse sampling process leads the frequency spectrum content of the digitised signal to be replicated around frequencies multiples of the sampling frequency (f_s), as exact duplicates of the original input signal spectrum. Restricting the minimum f_s to be at least two times the frequency range of the input signal allows to position one of these replicas in the DC to one half of f_s bandwidth with a minimum loss of information. Another source of loss of information rises from the uncertainty associated to the level where one code changes to an adjacent one. This fact requires the introduction of probability in the definition of ADC's models. Analogue-to-digital coder or analogue-to-digital encoder are synonymous often found of A/D converter.

Note:

- This quantisation procedure introduces inherent errors of 1/2 LSB (Lest Significant Bit) in the representation since, within the fractional range, only one (input) analogue value can be represented free of error by a single digital output code

- Physically an ADC may be a standalone device implanted in a single package, or a functional core (macro-block) embedded in a larger integrated system

- AC-coupled ADCs digitise only the AC component of the analogue input signal by blocking the static DC portion

See also: Analogue-to-digital conversion law (5.1), Quantisation (5.3), Transfer characteristic (5.5), Output coding (6), LSB (6.8), Full scale range (9.2).

5.3   Quantisation

The division of a quantity into a discrete number of small parts, often assumed to be integral multiples of a common quantity. It is a non-linear process in which the measured amplitude of an input signal at any instant is rounded off to the nearest of a set of predetermined values defined by the limits of non overlapping subranges. Whenever the signal value falls within a given subrange, the output has the corresponding discrete value [4]. In broad terms, quantisation is a non-linear operation that is carried out whenever a physical quantity is represented numerically by an integer corresponding to the nearest whole number of units. This process suggests that quantisation is like sampling in amplitude, a sampling process that acts not upon the function itself, but upon its probability density distribution (this definition does not apply to ADCs with triangular channel profile).

5.4   Straight line

Ideal straight line In an ideal ADC transfer characteristic, a straight line between the specified points for the most-positive (least-negative) and most-negative (least-positive) nominal code transition or code midstep values, respectively (see figures 1.5, 1.6)[2].

Fitting straight line In a real ADC transfer characteristic, a straight line through the measured output codes which fits the transfer characteristic according a specified criteria:

- End points straight line — the fitting line connecting the two end code transition, or the two end code midstep, values

---

[2]The ideal straight line passes through all the points for nominal code transition or code midstep values, respectively.

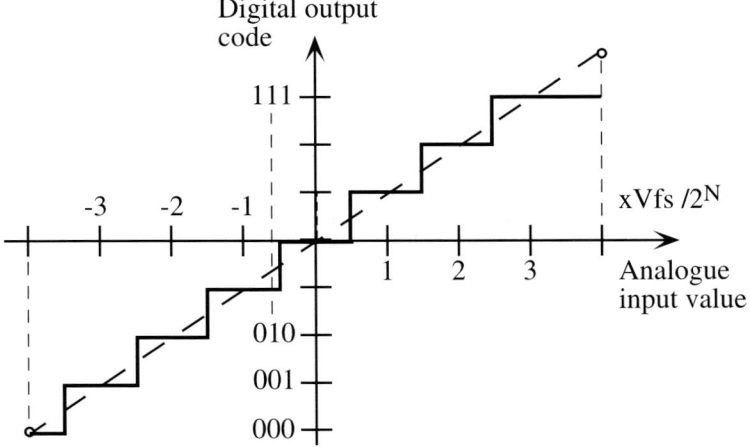

Figure 1.4. Typical transfer characteristic for a true-zero, binary coded, bipolar converter.

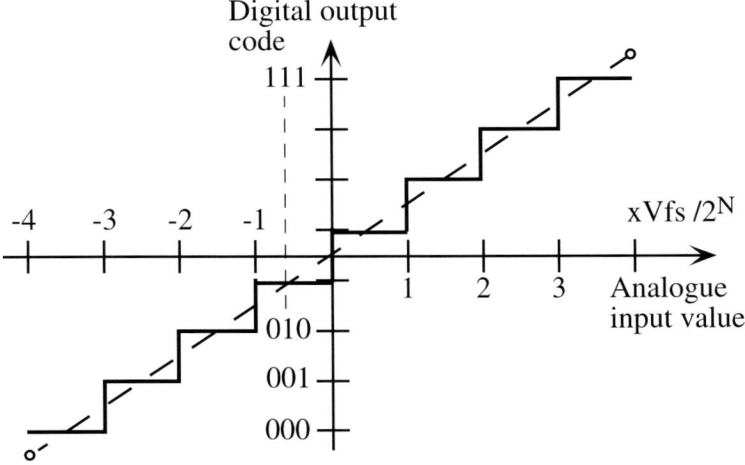

Figure 1.5. Typical transfer characteristic for a non-true-zero, binary coded, bipolar converter.

- Least-square fit straight line — the line which best approximates the measured transfer characteristic according a least-square fitting algorithm

- Minimum-maximum straight line[3] — the line which leads to the most positive and the most negative deviations from the ideal straight line to be equal in amplitude

---

[3]The minimum-maximum straight line tends to fall in disuse, however its definition is left here.

5.5   Transfer characteristic

A graphical staircase representation (see figure 1.6) of a deterministic conversion law, where the abscissa reports, either in absolute units or rationed to the full scale V_{fs}, the analogue input x, while the corresponding digital output codes are reported, at a constant pace, along the vertical axis. For each code, at the corresponding ordinate, a horizontal segment marks the interval(s) of abscissa values (code bin analogue input range) which are mapped to that code. The vertical segments of the staircase are in correspondence of the code transition levels. Mathematically, the transfer characteristic is a set of correlations between each of the fractional parts of the total analogue input range and the corresponding digital output codes.

5.5.1   Nominal transfer characteristic.   The transfer characteristic determined by the nominal code transition levels. A typical nominal transfer characteristic for a binary coded unipolar converter is shown in figure 1.6. The nominal code transition levels are placed at abscissas $T_{nom}[k] = (k - 1/2)V_{fs}2^{-N}$, where $k = 1, 2, ...(2^N - 1)$. Two typical transfer characteristics for bipolar converters are shown in figures 1.4 and 1.5. The transition levels in a real ADC differ from the nominal ones; in some of them (adjustable ADCs) it is possible, by external trimming, to bring some of the transition levels (typically the first and the last) close to the nominal ones.

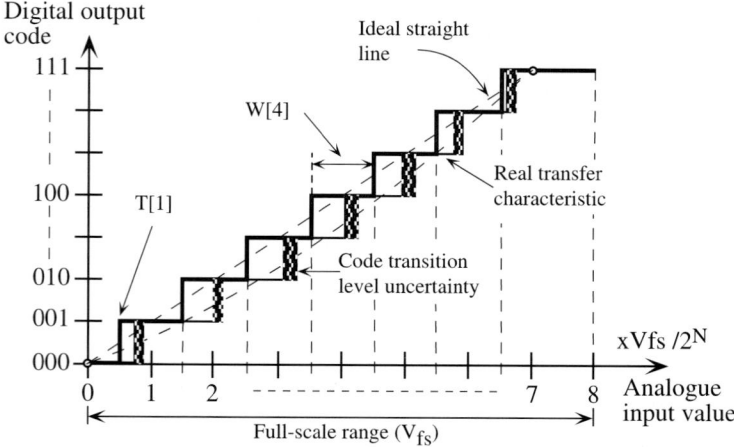

Figure 1.6.   Typical transfer characteristic for a binary coded unipolar converter.

5.5.2   Average transfer characteristic.   The representation of the average ADC output code, $\bar{y}$, as a function of the input signal value x.

See also: Deterministic conversion law (5.1.1), Average output code (6.1), Code bin (6.3), Code transition level (6.5).

6. Output coding

6.1 Average output code

For a given value of input $x(t)$, it is an estimate of the expected value $E[y[k]]$ of the output code $y[k]$. The arithmetic mean of M samples of $y[k]$ is used as an estimator of the output code.

6.2 Bandwidth (BW)

The band of frequencies of the input signal that the ADC under test is intended to digitise with nominal constant gain. It is also designated as the half-power bandwidth, i.e., the frequency range over which the ADC maintains a dynamic gain level of at least $-3 \, dB$ with respect to the maximum level.

See also: Gain flatness error (7.5).

6.3 Code bin or quantisation cell (k)

A code bin represents both, a fractional range of the analogue input quantity, and the correlated digital output code (see figure 1.7). Thus, a code bin can be defined, in a deterministic quantisation law, either as a digital output code $y[k]$ that corresponds to a particular set of input values x, each of which is uniquely mapped to that code, or as the interval (intervals) of values of x which correspond to the same output code k. The term step instead of code bin is often used whenever a qualitative rather then quantitative description is done.

See also: Code bin width (6.4), Quantisation (5.3).

6.4 Code bin width ($W[k]$)

In a monotonic ADC, the absolute value of the difference between the two ends of the range of analogue input values corresponding to one code bin (figure 1.6), i.e., the code transition levels $T[k]$ and $T[k+1]$ that delimit the kth bin:

$$W[k] = T[k+1] - T[k] \tag{1.1}$$

Nominal code bin width (Q) In an ideal ADC, the difference between the last and the first nominal code transition levels, divided by the total number of code bins encompassed between the two in the nominal conversion characteristic. In a N-bit, binary, unipolar ADC:

$$Q = \frac{T_{nom}[2^N - 1] - T_{nom}[1]}{2^N - 2} \tag{1.2}$$

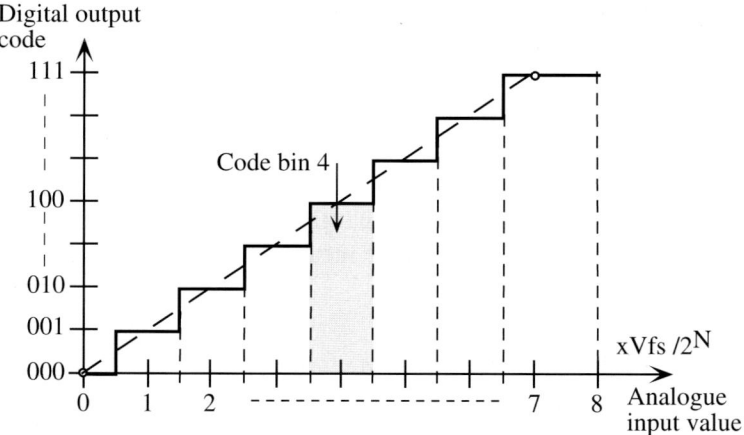

Figure 1.7.   Code bin representation.

NOTES:

- Different code bin widths along the ADC transfer characteristic result in a differential non-linearity error (DNL).

- The term step width instead of code bin width may be used whenever a qualitative rather than quantitative description is done.

See also: Quantisation (5.3), Code bin (6.3), Code transition level (6.5), Monotonic ADC (6.10), DNL (7.9.2).

6.5   Code transition level ($T[k]$)

In a deterministic conversion law, the analogue input value at which the change between two adjacent digital output codes takes place. Two consecutive code transition levels ($T[k]$, $T[k+1]$) are the extremes of an interval defining a code bin. A N-bit deterministic ADC presents $2^N - 1$ code transition levels (figure 1.6).

According to the considerations in the definition of A/D probabilistic conversion law, a more operational definition considers a transition level as the value of the converter input that causes half of the digital output codes to be greater than or equal to k, and half less than k (see figure 1.3). It follows that the transition $T[k]$ between codes ($k-1$) and k may occur at several different input levels.

Nominal code transition level ($T_{nom}[k]$) The code transition level specified by the manufacturer as a function of V_{fs} [4].

---

[4] A deviation of the code transition levels from their respective nominal values leads to the occurrence of an integral linearity error in the transfer characteristic.

Midcode bin value The analogue value for the centre of the code bin, except for the code bins at the two ends of the total range of analogue values[5].

Nominal midcode bin value A specified analogue value within a code bin that is ideally represented free of error by the corresponding digital output code.

Alternation band A property of some ADCs which identifies the input quantity range whereby the output alternates between two adjacent digital output codes.

See also: Conversion law (5.1), Code bin (6.3), Hysteresis (7.7), Integral non-linearity (7.9.1), Coherent sampling (8.2).

6.6 Effective Number of Bits (N_{ef})

In a first approach the nominal resolution of an ADC is determined by its physical number of bits. Actually the effective number of bits of an ADC identifies the resolution of the converter after correction for the signal-to-noise and distortion ratio. For a sinusoidal input signal, N_{ef} is given by:

$$N_{ef} = \frac{SINAD_{dBFS} - 1.76dB}{6.02} \tag{1.3}$$

where, $SINAD_{dBFS} = SINAD_{dB} - 20\log(SFDR)$.

The effective number of bits is a global indication of the real ADC's accuracy in terms of number of bits at a specific input frequency and sampling rate. In an actual ADC the measured error is generated by quantisation noise together with other sources of noise such as jitter, non-linearities, fixed pattern noise, reference voltage noise, power supply noise, missing codes, and thermal noise. The N_{ef} degrades with an increasing of the sampling frequency and its worst-case value occurs at the maximum specified sampling frequency ($f_{s_{max}}$) for an input frequency close or above one half of $f_{s_{max}}$.

See also: SINAD (7.19), SFDR (7.21).

6.7 Gain

Gain (G) The slope of the fitting straight line of the transfer characteristic, or of a specified part of it, expressed as the quotient of a change in digital output quantity (stated in the input quantity dimensions) by the change in analogue input quantity producing it .

---

[5]For the end codes, the midcode value is defined as the analogue value that results when the analogue value for the transition to the adjacent code is reduced or enlarged as appropriate by half the nominal value of the code bin width (see figure 1.6).

Gain point A point in the transfer characteristic corresponding to the code transition, or code midstep, value for which a gain error is specified, and in reference to which a gain adjustment[6] is performed (see figure 1.8).

See also: Gain and offset error (7.4).

6.8   Least-significant bit (LSB)

When referring to the input signal amplitude of a linear ADC, an LSB is synonymous with one ideal code bin width Q, which serves as a reference unit to express the magnitude of other analogue quantities of that same converter, especially of analogue errors, as multiples or submultiples of the analogue resolution. When describing the output bits of an ADC, LSB identifies the bit that has the lowest positional weight in a natural binary numeral.

See also: Code bin width (6.4).

6.9   Linear ADC

An ADC having code steps ideally of equal width excluding the code steps at the two ends of the total range of analogue input values[7].

See also: Linearity error (7.9).

6.10   Monotonic ADC

An ADC whose transfer characteristic is a monotonic function, i.e., which presents a positive sign derivative. A monotonic ADC ensures that the increase or decrease of the digital output is consistent (disregarding random noise) with an increase or decrease of the analogue input. An intermediate increment or decrement with the value zero does not invalidate monotonicity.

See also: Transfer characteristic (5.5), Missing code (7.10).

6.11   Non-linear ADC

An ADC with a specified non-linear transfer characteristic between the nominal code midstep values and the corresponding code step widths[8].

See also: Linearity error (7.9).

---

[6]Gain adjustment causes only a change of the slope of the transfer characteristic straight line (as defined above), without changing the offset.
[7]Ideally, the width of each end code step is one half of the width of any other code step.
[8]The function may be continuously non-linear or piece-wise linear.

7. Errors, non-linearity, noise, and distortion

7.1 Code error rate (w)

Ratio of the number of erroneous output codes to the total number of output codes in a record of data of specified length M. This ratio identifies the probability of the ADC incorrectly converting an input. This measure assumes that correction was made for gain, offset, and linearity errors, and noise floor amplitude is known. Possible causes for the occurrence of erroneous output codes are missing codes, noise interference, crosstalk, timing jitter, and aperture uncertainty. A common equivalent term is word error rate.

See also: Missing code (7.10), Noise (7.11), Aperture uncertainty (8.1), Jitter (8.7).

7.2 Dynamic range (DR)

Dynamic range is usually expressed in dB and describes the range of input signal levels that can be reliably measured simultaneously, in particular the ability to accurately measure small signals in the presence of large signals. In this sense, dynamic range can be defined as the ratio of the ADC full scale input range to the amplitude of the highest harmonic or peak noise floor. For an N-bit ideal ADC its value is equal to the ideal SINAD ($DR = 10 \log(4) \times N + 10 \log(1.5)$) as the only noise present in the output is quantisation noise.

For audio applications, and for sinusoidal inputs, it is common to consider the smallest input amplitude that which results for the case of a 0 dB SINAD. In this case, dynamic range is thus defined as

$$DR = 10 \log \frac{P_{input\ signal}}{P_{input\ signal@SINAD=0}} \tag{1.4}$$

Dynamic range identifies the range of input signal amplitudes that can be reliably converted simultaneously to a specified accuracy, expressed as the maximum ratio of the two signal levels. Considering the presence of noise floor, the minimum amplitude of a signal present at the ADC input which allows to detect the presence of this signal in the ADC output spectrum, is restricted by the peak amplitude of the noise floor.

See also: SINAD (7.19), Full-scale range (9.2).

7.3 Fixed-pattern noise

Noise due to localised non-uniformities in the transfer characteristic of an ADC, e.g., a missing code, which may originate spurious tones in its output spectrum.

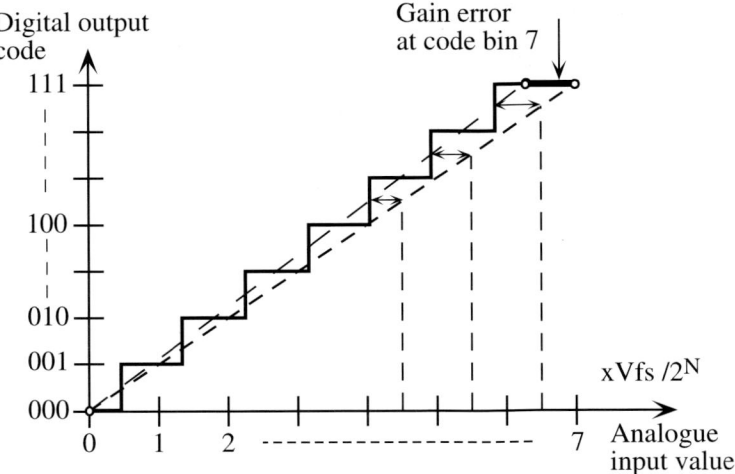

Figure 1.8. Gain error (after correction for the offset error).

7.4 Gain (G) and offset error (V_{os})

From code transition levels The values by which the specified code transition, or code midstep, values are multiplied and then to which the rescaled code transition, or code midstep, values are added, respectively:

- to cause the deviations from the nominal code transition, or code midstep, values to be zero at the terminal points, that is, the first and the last levels in the transfer characteristic — **end-points straight line**.

- to minimise the mean squared deviation of the actual code transition, or code midstep, values (measured fitting straight line) from the nominal code transition, or code midstep, values (ideal straight line) — **least squares fit definition**.

- to cause the most positive and the most negative deviation from the nominal transition levels to be equal in magnitude — **min-max definition**.

According to Mahoney [92] (page 119) using the two end points (only) to determine gain error should not be trusted for accurate slope measurement, especially in dynamic testing. The slope of the best fit line is a better indicator.

7.5 Gain flatness error ($\varepsilon_G(f)$)

The difference between the gain of the ADC at a given frequency in the ADC bandwidth, and its gain at a specified reference frequency, expressed as a percentage of the gain at the reference frequency. The reference frequency is typically the frequency where the bandwidth of ADC presents the maximum gain. For DC-coupled ADCs the reference frequency is usually $f_{ref} = 0$.

The gain flatness measures the variation of gain over a specified bandwidth, given by the maximum peak to peak gain variation at all output frequencies within this frequency range, under all specified conditions.

See also: Bandwidth (6.2).

7.6   Harmonic distortion

For a pure sine wave input , it is the set of output components at frequencies that are an integer multiple of the applied sine wave frequency f_1 up to the frequency $f_m = m \times f_1$, where m is application dependent. Harmonic distortion is due to the presence of frequencies in the output signal that are not seen in the input signal. It is caused by non-linearities within the device. As the importance of these harmonics in the output signal varies from application to application, the harmonic specification parameters should be calculated considering all the harmonics in the frequency range of interest for the application.

See also: Total harmonic distortion (7.23).

7.7   Hysteresis

The maximum difference in values of a code transition level, when the transition level is approached by a changing input signal from either side of the transition. The specified measurement conditions must include at least the shape and amplitude of the input signal and the sampling frequency[9].

See also: Code transition level (6.5), Coherent sampling (8.2).

7.8   Intermodulation distortion (IMD)

For an input signal composed of two or more pure sinewaves, output components at frequencies that are the sum and difference frequencies for all possible integer multiples of the input frequency tones. For example, for an input signal composed of two sinewaves, $f_1 = J_1 \frac{f_s}{M}$ and $f_2 = J_2 \frac{f_s}{M}$, the intermodulation distortion terms are those given by $f_{imf} = |jJ_1 \pm iJ_2| \times \frac{f_s}{M}$ where $i, j = 1, 2, 3, \cdots$, and J_1 and J_2 are chosen to be relative prime to M. IMD is given by the ratio of the *rms* sum of the individual distortion components to the *rms* amplitude of the input signal expressed in dB. Intermodulation distortion characterises non-linear distortion due to the appearance, in the output of the

---

[9]The presence of hysteresis phenomena is clearly inconsistent with the memory-less ADC model which is at the basis of both the probabilistic and deterministic conversion laws defined in, respectively, 1.5.1.2 and 1.5.1.1, and as a consequence with the definition chosen for code transition levels.

an ADC, of frequencies that are linear combinations of the fundamental frequencies and all harmonics that may eventually be present in the input signals. If input harmonic components are present they are not usually considered to characterise intermodulation distortion. When the harmonics are considered to be part of the distortion, a statement to that effect should be made.

7.9   Linearity error

7.9.1   Integral non-linearity ($INL[k]$).   The measure of the deviation of the code transition, or code midstep, levels from the ideal straight line, when an AC signal of specified shape, frequency (f), and amplitude (A), is converted at a specified sampling frequency (f_s) high enough to fully characterise the test frequency. The integral non-linearity error $INL[k]$ associated with code transition level k is the difference between the measured and nominal code transition levels, or the respective code midstep levels, after correcting the measured levels for gain and offset. It is usually expressed in LSB. When the code transition level is not specified, the absolute value of the maximum difference observed among all the possible values of k is reported. When the integral non-linearity is given as one number without code bin specification, it is the maximum integral non-linearity of the entire range.

Integral non-linearity is a measure of the ADC's accuracy. When spanning a converter over its full-scale range, its INL determines the deviation of its transfer characteristic from the expected ideal straight line of the converter. Some measures of INL take as a reference not the ideal straight line, but instead the straight line that best fits the transfer characteristic of the ADC. INL values calculated taking as a reference a best fit line are likely to be twice better than the values given by the end points measure. After correcting for offset and gain the end-points defined straight line is coincident with the ideal straight line. To ensure monotonicity $INL \leq 0.5$ LSB is required [20].

7.9.2   Differential non-linearity ($DNL[k]$).   The difference, after correcting for gain error, between the k-th code bin width $W[k]$ and the nominal code bin width Q, divided by the nominal code bin width, when a sinewave of specified frequency (f), and amplitude (A), is converted at a specified sampling frequency (f_s) high enough to fully characterise the test frequency. When the code k is not specified, the absolute value of the maximum difference observed out of all code bin widths deviations is reported.

Differential non-linearity is a measure of the linearity between code transitions. A DNL value different from zero identifies any deviation of the difference between two consecutive analogue code transition levels from the nominal code bin width.

7.9.3   Phase non-linearity.   A measure, in degrees, of the deviation of the ADC phase response from a straight line, as a function of frequency. It measures the ADC phase response that results when the rate of change of phase delay with input signal frequency over a specified bandwidth is not constant. This can cause signal distortion when different components of a signal arrive out of phase. It is usually expressed as one-half of the difference between the delays at the two extremes of the bandwidth being considered. If the worst case deviation value is provided, the frequency at which it occurs should be mentioned. It is also known as envelope delay distortion.

7.10   Missing code

An intermediate code that is absent when the changing analogue input to an ADC causes a multiple code change at the digital output. In general it can be defined as a code to which a differential non-linearity error $DNL[k] = -1$ is associated[10].

See also: Differential non-linearity (7.9.2).

7.11   Noise (total) (η)

Any deviation between the output signal (converted to input units) and the input signal, except deviations caused by linear time invariant system response (gain and phase shift) or a DC level shift. Noise is caused by either phase or amplitude noise like, quantization, thermal noise, fixed-pattern errors, non-linearities, aperture uncertainty, and jitter.

See also: Fixed-pattern noise (7.3), Random noise (7.17), Aperture un-certainty (8.1), Jitter (8.7).

7.12   Noise floor (η_f)

The spectrum of frequencies observed in an ADC output response, exclud-ing the fundamental frequency, its harmonics up to frequency f_m (being m application dependent), or DC level shift. This definition of noise comes from the fact that two parameters characterize the signal to noise ratio (SNR and SINAD) whose difference from each other relies on the separation between harmonic distortion and noise. All the harmonics from the fundamental up to frequency f_m (to be defined in each case) are taken into account for harmonic distortion evaluation. The remaining frequencies above the specified f_m are included in the broad classification of noise. Noise floor amplitude restricts the lowest input signal level which will produce a detectable digitally coded

---

[10]In [8] a DNL value smaller than -0.85 LSB is considered as giving origin to missing codes. An ADC having a DNL greater than 1 LSB is not guaranteed to have a missing code, though with all probability a missing code will occur.

equivalent information content at the output of the ADC. The noise floor limits the sensitivity to low level signals, since any signal with an amplitude smaller than the noise floor amplitude will result in an output signal with a $SNR \leq 1$ making it difficult, or even impossible to recover.

See also: Fixed-pattern noise (7.3), Random noise (7.17), Aperture uncertainty (8.1), Jitter (8.7).

7.13   Noise power ratio (NPR)

The ratio, expressed in dB, of the average power spectral density magnitude of an ADC output spectrum within a specified frequency range in response to a broad bandwidth signal, to the average power spectral density magnitude of the ADC output spectrum in the same frequency range when a specified frequency range is notch filtered from the broad bandwidth signal. The noise power ratio measures the difference in noise power measured at the output of the ADC under test, before and after inserting a bandstop notch filter between a broadband noise source and the ADC. The notch filter allows to remove selected bands of noise within which NPR is measured.

7.14   Offset

Offset The systematic deviation between ideal and actual code transition levels. It may be expressed in fractional LSB, or as the equivalent shift of the mean analogue input quantity required to eliminate the output deviation, expressed as an absolute value or as a percentage of full-scale range. Usually it is specified for code zero.

Offset point A point in the transfer characteristic corresponding to the code transition, or code midstep, value for which the offset error is specified, and by reference to which the offset adjustment must be performed[11].

See also: Gain and offset error (7.4).

7.15   Phase noise

Any fluctuation of the zero crossing instants of a waveform. It could include both long-term and short-term phase or frequency fluctuations. The long-term ones are generally specified in terms of frequency drift, being only the short-term ones those considered as phase noise. In the frequency domain phase

---

[11]Offset adjustment causes only a parallel displacement of the transfer characteristic, without changing its slope.

noise is expressed, in dB, as

$$\mathcal{L}(f)_{dB} = 20 \log[\frac{V_N(1H_z BW)}{V_C}] \qquad (1.5)$$

where, $V_N(1H_z \text{ BW})$ is the *rms* noise level in a 1 Hz bandwidth at a frequency separated from the waveform frequency (carrier) by f Hz (frequency offset), and V_C is the *rms* amplitude of the waveform.

See also: Jitter (section 8.7).

7.16   Quantisation noise / quantisation error

Noise caused by the error of approximating a variable having a continuous range of values to a quantised form having only discrete values. Quantisation noise depends on the particular analogue-to-digital conversion process used, and the statistical characteristics of the quantised signal. The inherent quantisation error of an ideal ADC (within a code), is the maximum (positive or negative) possible deviation of the actual analogue input value from the nominal midcode value. It can be shown [146] that this noise has the characteristics of white noise which has a rectangular probability distribution.
NOTES:

- This error is inherently due to quantisation. For a linear ideal ADC, its value equals ± 1/2 LSB (see figure 1.4).

- The term "resolution error" for the "inherent quantisation error" is deprecated, because "resolution" as a design parameter has a nominal value only.

See also: Quantisation (5.3).

7.17   Random noise

In general is an unpredictable (non-deterministic) disturbance interfering with a signal, described by its frequency spectrum and its amplitude statistical properties. In an ADC the disturbed signal may be either the input or the output signals.

7.18   Residuals

In a sine wave curve fitting procedure the residuals result from the difference between the recorded data and the fitted function.

7.19   Signal to noise and distortion ratio ($SINAD$)

For a pure sinewave input of specified amplitude and frequency, the ratio of the *rms* amplitude of the ADC output fundamental tone to the *rms* amplitude of the output noise, where noise is defined as to include not only random errors but also non-linear distortion and the effects of sampling time errors, i.e., the sum of all non-fundamental components in the range from DC (excluded) up to half the sampling frequency ($\frac{f_s}{2}$). In different words SINAD describes the quality of an ADC's dynamic range expressed as the ratio of the maximum amplitude output signal to the smallest increment of output signal that the converter can produce. In some glossaries it is also designated as Signal-to-THD plus noise.

See also: Noise (7.11), Signal to noise ratio (7.20), Spurious free dynamic range (7.21), Total harmonic distortion (7.23).

7.20   Signal to noise ratio (SNR)

Signal-to-noise ratio is a measure of the broadband noise and spurious that are introduced into the signal by the sampling and analogue-to-digital conversion processes. It is given by the ratio expressed in dB of the *rms* amplitude of the ADC output fundamental tone to the *rms* amplitude of the output noise, where noise is defined as the sum of all frequencies in the Nyquist band ($\frac{f_s}{2}$) excluding DC, fundamental, harmonics, and spurious components.

The SNR expresses the ratio of signal power to noise power in a specified frequency range. For an ideal ADC, i.e., which presents only a noise floor established by its quantisation process and no distortion, the SNR value is equal to SINAD. Assuming a sinusoidal input with a peak-to-peak amplitude equal to the full-scale range, the maximum achievable SNR is $SNR = 6.021N + 1.763 + 10log\frac{f_s}{2.f_{max}}$. Thus for sampling frequencies above the Nyquist rate of twice the maximum input frequency the SNR increases accordingly. This SNR improvement is often designated as "processing gain", and occurs because of a spreading of the quantisation noise power (assumed constant and independent of the bandwidth) as the sampling frequency is increased, that is, the noise that falls within the range from DC to f_{max} is minimised.

In the context of ADCs, the parameters SINAD (1.7.6) and SFDR (1.7.21) should be used instead.

See also: Dynamic range (7.2), Noise (7.11), Signal to noise and distortion ratio (7.19), Spurious free dynamic range (7.21), Total harmonic distortion (7.23).

7.21   Spurious-free dynamic range ($SFDR$)

SFDR expresses the range, in dB, of input signals lying between the averaged amplitude of the ADC's output fundamental tone, f_i, to the averaged amplitude of the highest frequency harmonic or spurious spectral component observed over the full Nyquist band, for a pure sinewave input of specified amplitude and frequency, i.e., $\max\{|Y(f_h)|, |Y(f_{sp})|\}$:

$$SFDR(dB) = 20\log\frac{|Y_{avg}(f_i)|}{max_{f_{sp},f_h}[|Y_{avg}(f_h)|\,,\,|X_{avg}(f_{sp})|]} \qquad (1.6)$$

where:

- Y_{avg} is the averaged spectrum of the ADC output,

- f_i is the input signal frequency,

- f_h and f_{sp} are the frequencies of the set of harmonic and spurious spectral components.

SFDR defines the difference in signal strength between the signal of interest and any other present in the band of interest. As generally harmonics limit SFDR, the worst case harmonic distortion is often specified to indicate SFDR.

7.22   Spurious tones

Undesired signal components usually at frequencies f_{sp} unrelated to the input signal frequency or its harmonic or intermodulation frequencies. One such frequency is considered a spurious tone if its spectral line amplitude is at least 10 dB higher than the noise floor in the averaged power spectrum.

7.23   Total harmonic distortion (*THD*)

The ratio of the rss (root-sum-of-squares) of all the harmonic distortion components, including their aliases in the spectral output of the ADC, to the *rms* amplitude of the output fundamental component, expressed in dB. The input stimulus is assumed a pure sinewave of specified amplitude and frequency.

$$THD(dB) = 20log\frac{\sqrt{\sum_{h=2}^{m} X(f_h)^2}}{X(1)} \qquad (1.7)$$

Unless otherwise specified, THD is estimated considering the second through the tenth harmonics, inclusive.

Usually the first three harmonics represent most of the ADC output distortion. However, it is common that the range of harmonics considered to compute THD is not specified in the data sheets, or that this range varies from manufacturer to manufacturer. If a single harmonic is considered for calculating *THD* then we have a Single Harmonic Distortion.

See also: Harmonic distortion (7.6).

7.24   Total spurious distortion (*TSD*)

The root of the sum of the powers of the spurious components in the range from DC (excluded) up to half the sampling frequency ($\frac{f_s}{2}$), expressed as a dB ratio to the *rms* amplitude of the output component at the input frequency, for a pure sinewave of specified amplitude and frequency stimulus.

See also: Spurious tones (7.22).

As the parameters described in this section express the dynamic performance in a specified frequency range and are dependent of the amplitude figure used for the signals under consideration, the bandwidth as well as the amplitude of the signal (if different from the full-scale range), harmonics, and noise should be specified together with the parameters' values.

8.   Data acquisition and processing

8.1   Aperture uncertainty

In a sample-and-hold or ADC, aperture is the time elapsing between the activation of the "Hold" control signal and the instant when the input signal is disconnected from the retention mechanism. Aperture uncertainty is the total time uncertainty spanning from the sampling / converting pulse instant, and the instant the input signal is actually sampled / converted, due to causes such as noise, signal amplitude dependent delay variation (as in a flash ADC), and temperature. It is often used interchangeably with aperture jitter. It is specified as a *rms* value which represents the standard deviation of the sample instant in time.

While the aperture time leads to a time error which renders a phase change of the ADC output signal, the aperture uncertainty limits the maximum input signal frequency that can be accurately converted. During the aperture uncertainty time the input signal amplitude should not change more than a value which does not leads to an error in the output code (usually $\pm 1/2$ LSB).

See also: Jitter (8.7).

8.2   Coherent sampling

The sampling of a periodic waveform such that the total number of samples (M) in the data record, correspond to an integer number of cycles (J) of the input waveform. If the data record comprises a number of sample sets, their respective end points are continuous. Coherent sampling prevents leakage and ensures that the acquired samples are not redundant. M and J should be relative prime numbers, being M a power of 2 to allow for the use of fast Fourier transform algorithms. Coherent sampling requires satisfying the following relationship:

$$J \times f_s = M \times f_i \qquad (1.8)$$

where f_s is the sampling frequency and f_i is the input waveform frequency.

See also: Sampling frequency (8.15).

8.3   Decimation

Act of collecting every Dth sample from an ADC output sequence of M samples, leading a decimated record to comprise M_d samples. When output decimation is used, the decimated sample rate, $f_d = \frac{f_s}{D}$, should be used for any equations relating sample rate to input frequency (e.g., for equivalent time sampling), but the actual ADC sample rate f_s should be quoted as the sample rate in the test results.

See also: Sampling frequency (8.15).

8.4   Equivalent-time sampling

The sampling of a periodic waveform such that consecutive samples are captured at different relative instants over various cycles, and assembled in order that the record of samples represents a single observation of the waveform. Equivalent-time sampling exploits the aliasing phenomenon to virtually increase the usable bandwidth of the sampling process.

8.5   Histogram, Distribution

8.5.1   Code histogram.   An ensemble reporting the total number of samples received in each code bin k, $H[k]$, as a function of k.

See also: Code bin (6.3), Cumulated code histogram (8.5.2).

8.5.2   Cumulated code histogram.   An histogram reporting the total number of samples $H_c[k]$ received in the bins corresponding to codes smaller or equal to k, as a function of k.

See also: Code bin (6.3), Code histogram (8.5.1).

8.5.3   Probability density function.   A function that defines how likely it is to find an ADC output code $y[k]$ between code transition levels $T[k]$ and $T[k+1]$. More meaningfully, for a continuous signal $x(t)$, it is the probability $p_x(\xi)d\xi$ of x being in the neighbourhood of ξ.

8.6   Incoherent sampling

The sampling of a waveform such that the coherent sampling condition is not meet.

See also: Coherent sampling (8.2).

8.7   Jitter (σ_j)

Any fluctuation in the time domain of the instant at which an event occurs. When applied to the sampling frequency, any modulation of the sampling clock time (random clock-to-clock timing errors) which leads the sampling frequency to deviate from its nominal value. The presence of jitter renders a stochastic change in a series of measurements of the code transition level between two adjacent digital output codes. It can be designated also as phase noise, and used interchangeably with aperture uncertainty.

Jitter is due to rapid, short-term, random fluctuations in the phase of a wave, caused by time-domain instabilities. It leads to short-term frequency variations in the clock (or input signal) frequency which appear as energy at frequencies other than the carrier. It is usually expressed in dB relative to carrier power (dBc) on a 1-Hz bandwidth, which is given by $\sigma_j = 10log[0.5 \times S(f)]$ where $S(f)$ is the spectral density of phase fluctuations, or as a *rms* deviation in a specified frequency from the carrier.

See also: Code transition level (6.5), Aperture uncertainty (8.1), Sampling frequency (8.15).

8.8   Processing gain (PG)

When applied to a window, the ratio of the output signal to noise floor ratio, to the input signal to noise ratio. It is the inverse of the normalised Equivalent Noise Bandwidth (ENBW). It is usually expressed in dB.

This processing gain identifies actually a signal processing gain, which is the ratio of the signal to noise ratio of the processed signal to the signal to noise ratio of the unprocessed signal. The processing gain concept can be applied in general to the signal gain, signal-to-noise ratio, signal shape, or other signal improvement obtained from the input to the output of the processing element.

See also: Window (8.17).

8.9   Record of data

The sequential collection of M samples acquired at the ADC output, which compose the vector $y[n]$, with $n = 0, 1, \ldots M$.

8.10   Record time length

The time length of a data record. For instance, for an M-sample record acquired at the uniform sampling period T_s, the record time length is MT_s.

8.11   Relatively prime

Two integers are relatively prime when their ratio is irreducible, i.e., their greatest common divisor is 1.

8.12   Root-mean-square (*rms*)

The square root of the average of the squares of all data values in a record.

8.13   Root-sum-square (rss)

The square root of the sum of the squares of all data values in a record.

8.14   Sampling

The process of converting a continuous-time signal into a signal defined only at discrete values of time.

8.15   Sampling frequency (f_s)

The frequency of the convert command. It determines the conversion rate (the number of conversions per unit time) performed by the ADC.

Although the Nyquist sampling theorem states that the sampling frequency must be at least twice the maximum frequency included in the input signal, sampling rates lower than twice the maximum input frequency can still allow for an exact reconstruction of the information content if the input signal is a bandpass one. This technique is designated as "undersampling" or "bandpass" sampling.

NOTES:

- The maximum conversion rate shall be specified for full resolution.

- The conversion rate is usually expressed as the number of conversions per second.

See also: Coherent sampling (8.2), Decimation (8.3).

8.16   Synchronous and asynchronous sampling

Synchronous sampling occurs when an input signal is sampled maintaining its phase locked with the ADC sampling frequency, otherwise it is said that the sampling is asynchronous.

See also: Sampling frequency (8.15).

8.17   Window

A set of coefficients to be multiplied by the corresponding samples in a time data record to modify their relative weights giving, usually, more emphasis to the samples in the centre of the record, in order to increase the accuracy of the parameters extracted from the time data record which characterise the signal. Generally, windowing is used when estimating frequency domain properties.

See also: Record of data (1.8.9).

9.   Input characteristics

9.1   Dither, or Source dither

Random noise added to the ADC input signal prior conversion , with the purpose of shaping the power density spectrum of the quantisation noise, moving its energy towards higher frequencies. Normally its amplitude is made smaller than $\frac{1}{2}LSB$, to prevent the increasing of noise floor amplitude. Although there is a slight increase in noise level, and hence a decrease of the signal-to-noise ratio, spectrally shaped dither can minimise this apparent increase. Noise is less objectionable than distortion, and allows low-level signals to be observed more clearly.

9.2   Full-scale range (V_{fs})

The difference between the maximum and the minimum convertible input values defined by the ideal straight line, according to the specified transfer characteristic, expressed as a function of reference voltage V_{ref}, $V_{fs} = V_{ref}(2^N - 1)/(2^N)$.

Nominal full-scale range (V_{fsn}) The total range in analogue values that theoretically can be coded with constant accuracy by the total number of steps. A typical reported nominal value of the full scale range for an N-bit binary unipolar converter (see figures 1.5 - 1.6) is:

$$V_{fsn} = \frac{2^N}{2^N - 2}(T_{nom}[2^N - 1] - T_{nom}[1]) = 2^N \times Q \qquad (1.9)$$

9.3   Full-scale signal (V_{ifs})

One whose peak-to-peak amplitude spans the entire range of input values recordable by the ADC under test.
See also: Full-scale range (9.2).

9.4   Input units

The measurement units for the physical quantity representing the input signal $x(t)$.

9.5   Reduced input range (V_{rir})

The ADC's input quantity range spanning between code transition levels $T[1]$ and $T[2^N - 1]$.

9.6   Signal to Full Scale Ratio ($SFSR$)

For a pure sinewave input of specified amplitude and frequency, the ratio of the ADC's output fundamental tone to the amplitude of a full scale sinewave at the same frequency.

9.7   Overdrive (V_{OD})

The magnitude of the difference between the positive/negative peaks of the input sinewave and the first/last code transition level of the transfer characteristic.

See also: Code transition level (6.5).

Chapter 2

SINEWAVE TEST SETUP

Pierre-Yves Roy
now with
EADS Defence and Security Systems SA
Defence and Communications Systems
Rue Jean-Pierre Timbaud - Montigny le Bretonneux
78063 Saint Quentin Yvelines Cedex, France
pierre-yves.roy@eads-telecom.com

Jacques Durand
now retired from
THALES
L'Orée de Corbeville, BP 56 91401 Orsay, France

1.   Test Setup description

All the dynamic test methods using sinewaves described in this book make use of a test setup of the type described in figure 2.1. A distinct advantage of sinewave testing is that it is relatively easy to evaluate the purity of a sinewave, for instance using a spectrum analyzer, and it is also easy to improve this purity by suitable filtering. In addition, high quality sinewave synthesizers are available on the market.

The ADC under test is stimulated with the sinewave provided by a high purity sinewave generator. Generally, the spectral purity of the generator alone is not adequate to the purpose of testing. In that case, a bandpass filter has to be inserted between the source and the ADC in order to reduce noise and/or harmonic distortion. In some cases, level adapters and unbalanced to balanced signal converters have to be added. All this extra electronics is preferably placed between the source and the bandpass filter, so that its contributions to distortion and noise are filtered out. Finally, an impedance matching network is frequently required between the bandpass filter and the ADC.

For intermodulation distortion (IMD) testing, figure 2.2 describes the test setup for two-tone measurements. If additional tones are needed, the same

D. Dallet and J. Machado da Silva, (eds.), Dynamic Characterisation of
Analogue-to-Digital Converters, 47–60.

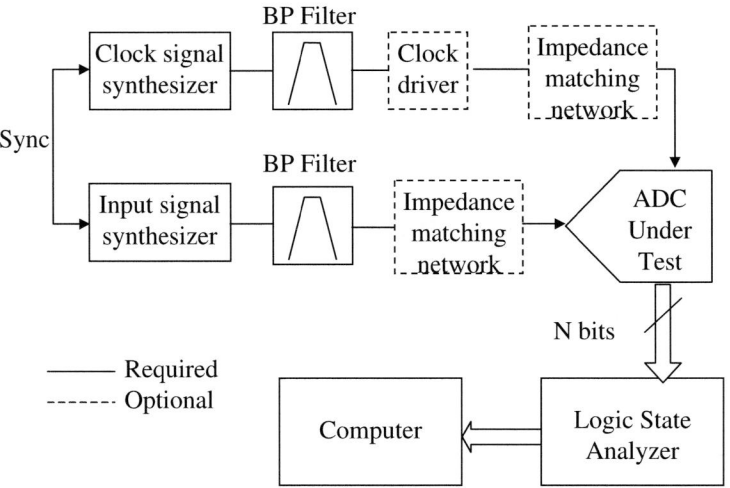

Figure 2.1. Test Setup for Sine Wave Testing.

type of setup can be used by combining the output of the required number of sinewave synthesizers. For IMD testing, an additional bandpass filter may be required after the combination of the tones in order to reduce the IMD of the input signal. As for single tone testing, an impedance matching network is frequently required between the bandpass filter and the ADC. The sampling clock is preferably derived from a second sinewave synthesizer: if this is indeed the case, the two synthesizers can be phase-locked, to guarantee that the sampling instants are placed in precise phase relationship with the input sinewave. This offers several advantages for the subsequent processing of the data, in particular by eliminating beat patterns which may render the measurement results unreliable.

The filter before the comparator/driver has the purpose of removing additive noise superposed to the sinewave, so as to reduce jitter at the output of the clock driver. A relevant advantage of using a sinewave generator to deliver the clock signal is related to the quieter EMC environment of the test bench: the high frequency harmonic contents related to the presence of a digital clock signal are confined to a small portion of the test board, between the comparator/clock driver and the ADC under test. Frequently it may be advisable to smooth the edges of the clock signal, to avoid leakages to other sensitive parts of the test board.

Note that an external frequency divider may be inserted in the clock chain, with the aim of achieving more closely the desired frequency ratio and/or reducing the phase noise of the sampling signal.

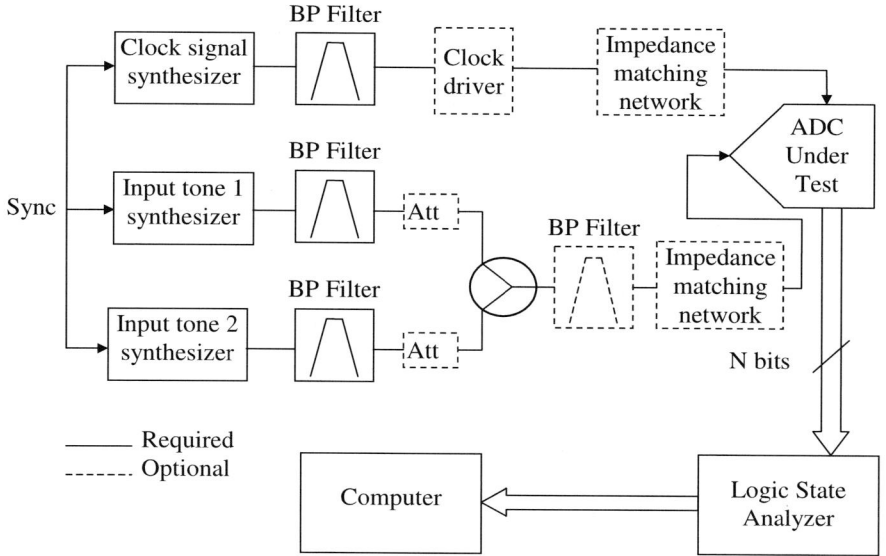

Figure 2.2. Test Setup for two-tone IMD Testing.

2. Specification of the clock and input signal

2.1 Harmonic distortion considerations

2.1.1 THD measurements.

The harmonic distortion of the input sinewave must be related to the THD of the ADC under test in order to minimize the errors in the measurements of the THD and also of the SINAD and/or of the SFDR.

In the worst case, the THD of the ADC under test and the input sinewave distortion are dominated by the same harmonic component (same frequency and same phase). In the case of one dominant harmonic component, the THD of the ADC can be approximated by

$$THD_{ADC} = \frac{A_{H_{ADC}}}{A}$$

$$THD_{ADC_{dB}} = 20log\frac{A_{H_{ADC}}}{A} \tag{2.1}$$

where $A_{H_{ADC}}$ is the amplitude of the dominant harmonic component created by the ADC and A is the amplitude of the fundamental component.

The THD of the input sinewave can be defined by

$$THD_{input} = \frac{A_{H_{input}}}{A}$$

$$THD_{input_{dB}} = 20log\frac{A_{H_{input}}}{A} \tag{2.2}$$

where $A_{H_{input}}$ is the dominant harmonic component of the input sinewave and A is the amplitude of the input sinewave.

As the dominant harmonic component created by the ADC and the input harmonic distortion have the same frequency and the same phase, the measured THD is

$$
\begin{aligned}
THD_{measured_{dB}} &= 20log\frac{A_{H_{ADC}} + A_{H_{input}}}{A} \\
&= 20log\left[\frac{A_{H_{ADC}}}{A}\left(1 + \frac{A_{H_{input}}}{A_{H_{ADC}}}\right)\right] \\
&= THD_{ADC_{dB}} + 20log\left(1 + \frac{THD_{input}}{THD_{ADC}}\right)
\end{aligned}
\tag{2.3}
$$

In that case, the error on the measurement of the THD is

$$20log\left(1 + \frac{THD_{input}}{THD_{ADC}}\right)$$

For an error lower than 0.5 dB on the measurement of the THD, the harmonic distortion of the input sinewave must be at least 25 dB better than the THD of the ADC.

In all the other cases, the THD of the ADC as well as the distortion of the input sinewave result from a distortion over many harmonic components. These components have not necessary the same frequency and/or the same phase. In that case, the THD of the ADC is defined by

$$THD_{ADC} = \frac{\sqrt{\sum_i A^2_{H_{i_{ADC}}}}}{A}$$

$$THD_{ADC_{dB}} = 20log\frac{\sqrt{\sum_i A^2_{H_{i_{ADC}}}}}{A} \tag{2.4}$$

where $A^2_{H_{i_{ADC}}}$ is the power of the ith harmonic component created by the ADC and A is the amplitude of the fundamental component.

The THD of the input sinewave can be defined by

$$THD_{input} = \frac{\sqrt{\sum_i A^2_{H_{i_{input}}}}}{A}$$

$$THD_{input_{dB}} = 20log\frac{\sqrt{\sum_i A^2_{H_{i_{input}}}}}{A}$$

(2.5)

where $A^2_{H_{i_{input}}}$ and A are respectively the power of the ith harmonic component and the amplitude of the fundamental component of the input sinewave.

In that case, the powers of the harmonic components must be added and the measured THD is

$$THD_{measured_{dB}} = 20log\frac{\sqrt{\sum_i A^2_{H_{i_{ADC}}} + \sum_i A^2_{H_{i_{input}}}}}{A}$$

$$= 10log\left[\frac{\sum_i A^2_{H_{i_{ADC}}}}{A^2}\left(1 + \frac{\sum_i A^2_{H_{i_{input}}}}{\sum_i A^2_{H_{i_{ADC}}}}\right)\right]$$

(2.6)

$$= THD_{ADC_{dB}} + 20log\sqrt{1 + \frac{THD^2_{input}}{THD^2_{ADC}}}$$

In the cases where the THD of the ADC and the distortion of the input sinewave result from a distortion over many harmonic components, the error on the measurement of the THD is

$$20log\sqrt{1 + \frac{THD^2_{input}}{THD^2_{ADC}}}$$

For an error lower than 0.5 dB on the measurement of the THD, the harmonic distortion of the input sinewave must be at least 9 dB better than the THD of the ADC.

Given the case considered and knowing the harmonic distortion of the input sinewave synthesizer, the rejection of the bandpass filter inserted between the source and the ADC can be calculated.

2.1.2   IMD measurements.   If the IMD of the ADC and the IMD of the input signal are dominated by the same intermodulation tone (IM tone) (same frequency and same phase), the same reasoning than the one used for the worst case THD measurement leads to an error on the measured IMD equal to

$$\epsilon_{IMD} = -20log\left(1 + \frac{IMD_{ADC}}{IMD_{input}}\right)$$

(2.7)

Where

$$IMD_{ADC} = \frac{A}{A_{IMtone_{ADC}}} \qquad (2.8)$$

and

$$IMD_{input} = \frac{A}{A_{IMtone_{input}}} \qquad (2.9)$$

with $A_{IMtone_{ADC}}$ is the amplitude of the dominant IM tone created by the ADC, A the amplitude of the lower input tone, and $A_{IMtone_{input}}$ the amplitude of the dominant IM tone of the input signal.

For an error lower than 0.5 dB on the measurement of the IMD, the intermodulation distortion of the input signal must be at least 25 dB better than the one of the ADC.

If the IMD of the ADC and the IMD of the input signal are not dominated by the same IM tone, the difference between the IMD of the input signal and the one of the ADC can be lower than 25 dB for an error of 0.5 dB on the measurement of the IMD of the ADC. This difference depends on the configuration of the IM tones at the input of the ADC and on the configuration of the IM tones created by the ADC, that is why it is impossible to give a general rule in that case. Nevertheless, the error given in (2.7) is a worst case one and for a given value of the error, the ratio between the IMD of the input signal and the IMD of the ADC calculated from this equation will ensure that, in any case, the IMD is always measured with the accuracy wanted.

2.2   Jitter considerations

The phase noise of the signal and clock sources produce errors for many dynamic test methods. The phase noise of a signal source is usually described by the SSB phase noise spectral power density $\mathcal{L}(f_o)$ (measured in dBc/Hz) as a function of the offset f_o from the carrier (in Hz), see figure 2.3. The variance (power) of the phase noise is calculated by[1]

$$\sigma_\theta^2 = 2 \int_{f_L}^{f_H} \mathcal{L}(f_o) df_o \qquad (2.10)$$

where f_L and f_H depend of the application in which the generator is used.

In the setup described in figure 2.1, a selective bandpass filter is placed between the ADC and the generator for the signal and the clock sources. If the filters' 3 dB bandwidth are $BW_{f_{in}}$ for the input signal and BW_{f_s} for the clock signal, the upper bound of the integral in (2.10) is $f_H = \frac{BW_{f_{in}}}{2}$ for the input signal and $f_H = \frac{BW_{f_s}}{2}$ for the clock signal.

---

[1] see equations (22.9) and (22.10) in [38].

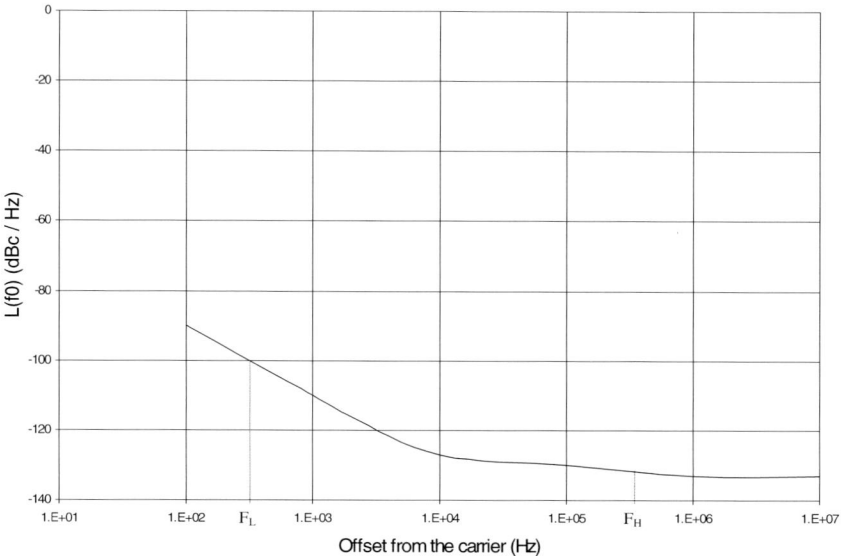

Figure 2.3. Typical SSB phase noise power density of a sinewave generator.

If we consider the acquisition of M samples at the sampling frequency f_s, the lower bound of the integral in (2.10) is $f_L = \frac{f_s}{M}$ for both signals.

So for the test setup described in figure 2.1, the variances of the phase noise of the input signal and of the clock signal can be calculated by

$$\sigma^2_{\theta_{sig}} = 2 \int_{\frac{f_s}{M}}^{\frac{BW_{f_{in}}}{2}} \mathcal{L}_{sig}(f_o)df_o$$

$$\sigma^2_{\theta_{clk}} = 2 \int_{\frac{f_s}{M}}^{\frac{BW_{f_s}}{2}} \mathcal{L}_{clk}(f_o)df_o$$

(2.11)

Note that if $\frac{BW_{f_{in}}}{2}$ (respectively $\frac{BW_{f_s}}{2}$) is lower than or equal to $\frac{f_s}{M}$, no phase noise is added to the input (respectively clock) signal.

The timing jitter variance is related to the phase noise variance by

$$\sigma^2_{T_{sig}} = \frac{\sigma^2_{\theta_{sig}}}{(2\pi f_{in})^2}$$

$$\sigma^2_{T_{clk}} = \frac{\sigma^2_{\theta_{clk}}}{(2\pi f_s)^2}$$

(2.12)

Table 2.1. Error (in dB) on SNR measurement due to timing jitter(1/2)

N $\downarrow$ \ $f_{in}\sigma_T$ $\rightarrow$	5E-4	1E-4	5E-5	1E-5	5E-6	1E-6
6	0.3	< 0.1	< 0.1	< 0.1	< 0.1	< 0.1
7	0.9	< 0.1	< 0.1	< 0.1	< 0.1	< 0.1
8	2.9	0.2	< 0.1	< 0.1	< 0.1	< 0.1
9	6.9	0.6	0.2	< 0.1	< 0.1	< 0.1
10	> 10	2.1	0.6	< 0.1	< 0.1	< 0.1
11	> 10	5.4	2.1	0.1	< 0.1	< 0.1
12	> 10	> 10	5.4	0.4	0.1	< 0.1
13	> 10	> 10	> 10	1.5	0.4	< 0.1
14	> 10	> 10	> 10	4.1	1.5	0.1
15	> 10	> 10	> 10	8.7	4.1	0.3
16	> 10	> 10	> 10	> 10	8.7	1.0
17	> 10	> 10	> 10	> 10	> 10	3.0
18	> 10	> 10	> 10	> 10	> 10	7.0

The variance of the total timing jitter of the test setup is

$$\sigma_T^2 = \sigma_{T_{sig}}^2 + \sigma_{T_{clk}}^2 \tag{2.13}$$

Note that the contribution of the clock driver to the total timing jitter has been neglected in (2.13). For most high speed ADCs, no external clock driver is required and (2.13) is valid. When an external clock driver is required, the phase noise of the clock signal must be measured after the clock driver in order to take into account its effects on the phase noise of the clock signal. In that case, $\sigma_{T_{clk}}^2$ is replaced by $\sigma_{T_{clk+driver}}^2$ in (2.13). Nevertheless, the contribution of the external clock driver to the total timing jitter is often negligible and (2.13) can be used.

The total timing jitter, causes a non-uniform sampling of the input signal, which results in an error on the measurement of dynamic parameters like SNR, SINAD, DNL, and INL.

For an ideal N-bit ADC, the SNR is

$$SNR_{ideal} = 10log(2^{2N-1} \times 3) = 6.02N + 1.76 \tag{2.14}$$

If the SNR of this N-bit ideal ADC is measured with a non-ideal test setup, the presence of timing jitter leads the measured value to be lower than the value calculated by (2.14). An expression of the measured SNR in presence of a timing jitter with a variance σ_T^2 is found in [130]

$$SNR_{measured} = -10log \left(\frac{1}{3 \times 2^{2N-1}} + (2\pi f_{in}\sigma_T)^2 \right) \tag{2.15}$$

Table 2.2. Error (in dB) on the SNR measurement due to timing jitter (2/2)

N ↓ \ $f_{in}\sigma_T$ →	5E-7	1E-7	5E-8	1E-8	5E-9	1E-9
15	0.1	< 0.1	< 0.1	< 0.1	< 0.1	< 0.1
16	0.3	< 0.1	< 0.1	< 0.1	< 0.1	< 0.1
17	1.0	< 0.1	< 0.1	< 0.1	< 0.1	< 0.1
18	3.0	0.2	< 0.1	< 0.1	< 0.1	< 0.1
19	7.0	0.7	0.2	< 0.1	< 0.1	< 0.1
20	> 10	2.2	0.7	< 0.1	< 0.1	< 0.1
21	> 10	5.6	2.2	0.1	< 0.1	< 0.1
22	> 10	> 10	5.6	0.4	0.1	< 0.1
23	> 10	> 10	> 10	1.5	0.4	< 0.1
24	> 10	> 10	> 10	4.3	1.5	0.1

The error on the SNR measurement due to the timing jitter of the test setup is

$$\epsilon_{SNR} = 6.02N + 1.76 + 10log\left(\frac{1}{3 \times 2^{2N-1}} + (2\pi f_{in}\sigma_T)^2\right) \qquad (2.16)$$

The SNR error measurement expressed in dB is listed in Tables 2.1 and 2.2, for diferent error ranges, as a function of N and $f_{in}\sigma_T$. It can be seen that very large errors can occur when the timing jitter of the test setup is not low enough.

The error calculated in the tables can also be expressed as

$$\epsilon_{SNR} = 6.02N + 1.76$$

$$+ 10log\left(\frac{1}{3 \times 2^{2N-1}} + \sigma^2_{\phi_{sig}} + \left(\frac{f_{in}}{f_s}\right)^2 \sigma^2_{\phi_{clk}}\right) \qquad (2.17)$$

For a given ADC, the only way to reduce this error is to lower the variance of the phase noise of the input signal and of the clock signal. In (2.17) the variance of the clock phase noise is multiplied by $\left(\frac{f_{in}}{f_s}\right)^2$. That reduces the effect of the clock phase noise for the measurements performed with an input signal in the 2 first Nyquist zones and increases its effect otherwise.

The value of the variance of the phase noise results from the integration of the SSB phase noise spectral power density of the generator (see (2.11)). The lower bound of the integral ($\frac{f_s}{M}$) is defined by the ADC under test. The upper one is half the bandwidth of the bandpass filter, that is why very narrow bandpass filters allow the power of the phase noise to be reduced.

Knowing the SSB phase noise spectral power density of generators and the input and clock frequencies, the bandwidth of the filters of the setup described in figure 2.1 can be specified in order to measure the SNR of the ADC with the accuracy wanted.

Using filters with a bandwidth such that $\frac{BW_{fs}}{2} \leq \frac{f_s}{M}$ and $\frac{BW_{fin}}{2} \leq \frac{f_s}{M}$ provides a reduced timing jitter test setup which does not affect SNR measurement errors. Unfortunately, it is often impossible to reach such a narrow bandwidth or it requires very expensive filters. However, thanks to the performances of the high purity sinewave generators available on the market, it is often not necessary to use bandpass filters with such narrow bandwidths.

For a given ADC, in order to reach the required accuracy on the SNR measurement, the maximum tolerable total timing jitter value is given in Tables 2.1 and 2.2. After that value, the generator's SSB phase noise spectral power density, and input and clock frequencies, the clock and signal bandwidths $((BW_{clk})$ and $BW_{sig})$ can be determined. It is very important to specify precisely the filters' bandwidths and not to overestimate them, because they often determine the filters' technology and thus there cost.

3.   Example of filter specification

Let's consider a 12-bit 50 MS/s ADC to be measured with an accuracy of 0.1 dB on the SNR and 0.5 dB on the THD. The typical value of the SNR given by the manufacturer is 68 dBFS for $f_{in} = 10MHz$, the typical THD at the same frequency is -80 dBFS. The record size for the characterization is 16384 samples. The synthesizers used for the clock and input signals are identical and their SSB phase noise power density is given in table 2.3.

Table 2.3.   SSB phase noise power density of the clock and signal generators.

Offset from the carrier (Hz)	$\mathcal{L}(f_o)$ (dBc/Hz)
1	-78
10	-108
100	-126
1000	-132
3000	-135
5000	-138
10000	-138
100000	-139

The highest harmonic generated by the synthesizer is the second one and its amplitude is 30 dB below the carrier. Let's consider that the THD of the ADC is also determined by the second harmonic. As described in 2.1, the THD of the input signal, being THD$_{ADC}$=-80 dBFS, must be at least -105 dB to reach the desired accuracy. That leads to specify the minimum rejection of the signal BP filter to -75 dB at 20 MHz. To get a maximum SNR measurement error of 0.1

dB with this ADC (68 dBFS), line N=12 of table 2.1 gives a maximum value for $f_{in}\sigma_T$ of 5.10^{-6}, that is to say a maximum timing jitter of $\sigma_T \leq 0.5ps$.

The bandwidth of the clock and signal band-pass filters must then be specified to limit the total jitter to $0.5ps$. As the signal synthesizers are the same for the clock and input signals, and as the clock frequency is five times the input signal frequency, the bandwidth of the clock filter can be wider than the one of the signal filter. Even if there are many bandwidth combinations that can lead to the correct setup, the more apropriate combinations of filter bandwidths are the ones that allow the lowest cost filters to be used. In this case, assuming that the timing jitter induced by the signal is 0.92 ps and the jitter induced by the clock is 0.4 ps ($sqrt(0.92^2 + 0.4^2) = 1$), a good bandwidth specification could be, according 2.11 and 2.12,

- 3 dB BW of the signal filter — $BW_{f_{in}} = 56,2\,kHz\,(0.6\%)$

- 3 dB BW of the clock filter — $BW_{f_s} = 711\,kHz\,(1.4\%)$

The percentage figures between brackets give the ratio between the bandwidth of the filter and its central frequency, which is sometimes called the relative bandwidth of the filter. When specified as described above, the two filters can be manufactured with a low cost technology (LC).

Another filters' combination still limiting jitter to 0.5 ps could be

- 3 dB BW of the signal filter — $BW_{f_{in}} = 24\,kHz\,(0.24\%)$

- 3 dB BW of the clock filter — $BW_{f_s} = 1520\,kHz\,(3\%)$

Now, the timing jitter induced by the signal is 0.24 ps and the clock induced jitter is 0.44 ps. This filters' specification leads to a much narrower signal filter, which might not be the best choice in terms of filters' costs.

4.   Filter selection

The explanations and tables given in section 2 allow the value of the rejection and of the bandwidth of the bandpass filters to be calculated. From these values, the feasibility and the technology adopted to design the filters can be determined. Table 2.4 lists the narrowest 3 dB bandwidths and the highest reachable rejections for filter technologies suitable for the frequencies commonly used in the dynamic testing of ADCs.

The relative 3 dB BW is the value of the 3 dB BW divided by the central frequency and the rejection is the difference between the stopband attenuation and the insertion loss.

The values in table 2.4 define rather standard filters for most of the very narrow and very high selectivity bandpass filter manufacturers. Nevertheless, some manufacturers can design specific filters with higher performances.

Table 2.4.   Filters performances

Technology	Center Frequency	Relative 3 dB BW	Rejection
L, C	DC → 1 GHz	1 %	70 dB
Quartz	10 kHz → 100 MHz	0.01 %	90 dB
Surface Acoustic Wave (SAW)	10 MHz → 2 GHz	0.02 %	60 dB
Tubular	50 MHz → 6 GHz	2 %	70 dB
Printed line	50 MHz → 18 GHz	3 %	60 dB
Dielectric Resonator	400 MHz → 3 GHz	1 %	60 dB

4.1   LC filters design

For the dynamic testing of ADC using sinewaves, band-pass filters are used in order to filter the noise and/or the harmonics. Generally, the parameters required for the filters used in sinewave test setups are a narrow 3 dB bandwidth and a high stop-band rejection. The most suitable filters to provide these characteristics are Chebyshev filters. Moreover, for sinewave testing, the ripple of that kind of filter is not a problem as it can be for wide-band signal testing.

The harmonic distortion and jitter considerations allow the value of the stop-band width, the stop-band attenuation and the 3 dB bandwidth to be determined. To design the filter, the input and the output impedances must also be known. Most of the time, with LC filters, these impedances equal 50 Ω and a impedance matching network is used to match the filter impedance to the ADC impedance.

Knowing the stop-band width and the stop-band attenuation, the order of the filter can be determined as shown in figure 2.4. The curves of figure 2.4 are reproduced from [150] and show the pass-band and stop-band attenuation as a function of the frequency normalized to the 3 dB bandwidth. Once the order of the filter is determined, [150] gives the normalized value of the inductors and capacitors to use to build the filter. The last step is to denormalize these coefficients using the central frequency, the 3 dB bandwidth and the input and the output impedances in order to get the real values of the components to use.

Today, this analysis is performed automatically with the filter synthesis softwares available on the market. Moreover, these softwares can take into account the quality factors of the inductors and capacitors and thus simulate the exact response curve of the filters that will be manufactured.

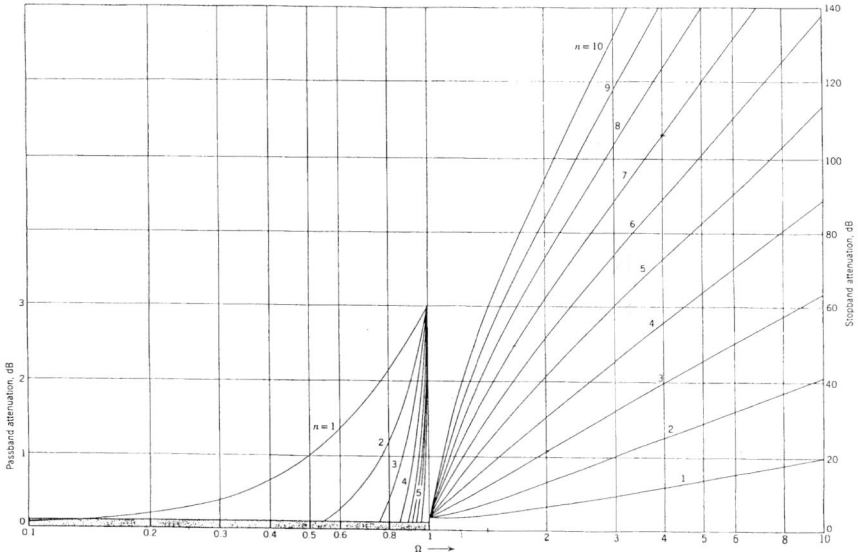

Figure 2.4. Attenuation characteristics for Chebychev filter with 0.1 dB ripple.

5.   Taking a record of data

A record of data is a sequential series of samples acquired by the logic state analyzer (or embedded processing element in the case of built-in self test) interfaced to the ADC. Once acquired, the record of data is transferred to a computer for analysis.

5.1   Use of Output Decimation in Taking a Record of Data

In the case of ADCs with very high sampling rate, it may become impractical to store in real time all the samples acquired by the ADC. In this case, it is still possible to operate the ADC at the high sampling rate, while storing only one sample out of D acquired in sequence (1/D output decimation). The set of samples thus collected takes the name of decimated record.

In order to acquire a decimated record, the ADC sampling clock is also used to drive a divide-by-d counter, whose output triggers the acquisition of the sample by the logic analyzer (or by the equivalent hardware).

In the case of output decimation, the decimated sample rate, f_s/D, shall be used in all the equations relating the sampling rate to the input frequency (e.g., for equivalent time sampling), but the actual ADC sampling rate f_s shall be quoted as the sampling rate in the test report. When a decimator factor D is applied, this implies a test time multiplied by D. So, in that case, the lower bound of the integrals used in (2.11) must be changed to $\frac{f_s}{M.D}$.

Output decimation may also be used for tests that do not require waveform reconstruction, such as histogram tests, or for servo-loop tests. In general, it should be remembered that decimation involves a loss of information, which may occasionally hide relevant phenomena, such as hysteresis effects.

Chapter 3

TIME-DOMAIN DATA ANALYSIS

Dominique Dallet
Laboratoire IXL-ENSEIRB, CNRS UMR 5818, Université Bordeaux 1
351 cours de la libération
33405 Talence Cedex, France
dallet@ixl.fr

Djamel Haddadi
STMicroelectronics
12, Rue Jules Horowitz
38019 Grenoble, France
djamel.haddadi@st.com

Philippe Marchegay
Laboratoire IXL-ENSEIRB, CNRS UMR 5818, Université Bordeaux 1
351 cours de la libération
33405 Talence Cedex, France
marchegay@ixl.fr

1.   Introduction

The time-domain analysis ADC testing method is based on the sinewave curve fitting algorithm, which uses linear or nonlinear least square method as function of the knowledge of the input frequency value. In this chapter algorithms to estimate the spectral parameters like $SINAD$, SNR, THD and $ENOB$ in time domain are proposed. Basic theorectical background is provided here, but for a more extended description of all mathematical details see [44], [62], [63], [64], [65].

D. Dallet and J. Machado da Silva, (eds.), Dynamic Characterisation of
Analogue-to-Digital Converters, 61–84.

2.   Calculation of the dynamic parameters

In order to measure A/D converter dynamic parameters, a clock signal of frequency f_s and a pure sinewave

$$x(t) = A\cos(2\pi f_i t + \phi) + C \tag{3.1}$$

are applied to the input of the ADC evaluation board. Then, M samples

$$\mathbf{y} = [y_0, \ldots, y_{M-1}]^T \tag{3.2}$$

are acquired. The objective is to evaluate the signal-to-noise and distortion ratio and the effective number of bits from the ADC output data by a sinewave fitting procedure.

3.   Definitions

The signal-to-noise and distortion ratio $(SINAD)$ is the power of the test signal to the power of noise and harmonics (see ADC terminology). It is generally expressed in dB and calculated using the equation

$$SINAD\ [dB] = 10\log_{10}\frac{(\frac{A^2}{2})}{\eta_{rms}^2} \tag{3.3}$$

where

$$\eta_{rms}^2 = \frac{1}{M}\sum_{n=0}^{M-1}(\mathbf{y}[n] - \mathbf{x}[n])^2 \ \text{ with } \mathbf{x}[n] = x(\frac{n}{f_s}) \tag{3.4}$$

The effective number of bits is given by

$$N_{ef}\ [Bits] = N - \log_2\frac{\eta_{rms}}{\sigma_{\varepsilon q}} \tag{3.5}$$

where N is the the ADC number of bits, and $\sigma_{\varepsilon q}$ is the ideal quantisation noise root mean square (rms). This latter is usually set equal to $\frac{Q}{\sqrt{12}}$, Q being the nominal code bin width.

Thus, the ADC parameters require the knowledge of the test sinewave parameters. These are latter estimated in the time-domain by minimising the least squares cost function

$$\chi(\mathbf{x}p) = \frac{1}{2}\sum_{n=0}^{M-1}(\mathbf{y}[n] - \mathbf{x}[n])^2 \tag{3.6}$$

with respect to the four parameters

$$\mathbf{x}p = [A\cos(\phi),\ A\sin(\phi),\ C,\ \omega_{in}]^T \tag{3.7}$$

where ω_{in} is the normalised angular frequency

$$\omega_{in} = \frac{2\pi f_i}{f_s} \tag{3.8}$$

We will distinguish two cases depending on whether ω_{in} is known or not.

4. The fixed-frequency method

If ω_{in} is known with a good accuracy, then we have only to estimate the three parameters

$$xp = [A\cos(\phi),\ A\sin(\phi),\ C]^T \tag{3.9}$$

The estimates of the linear parameters xp are determined by minimising the function $\chi(xp)$. This is done by noting that $\chi(xp)$ can be rewritten as

$$\chi(xp) = \frac{1}{2}\left[(\mathbf{y} - \mathbf{E}\hat{x}p)^T(\mathbf{y} - \mathbf{E}\hat{x}p) + (xp - \hat{x}p)^T\mathbf{E}^T\mathbf{E}(xp - \hat{x}p)\right] \tag{3.10}$$

where

$$\hat{x}p = (\mathbf{E}^T\mathbf{E})^{-1}\mathbf{E}^T\mathbf{y} \tag{3.11}$$

and $\mathbf{E}$ is the $M \times 3$ matrix

$$\mathbf{E} = \begin{bmatrix} 1 & 0 & 1 \\ \cos(\omega_{in}) & -\sin(\omega_{in}) & 1 \\ \cos(2\omega_{in}) & -\sin(2\omega_{in}) & 1 \\ \vdots & \vdots & \vdots \\ \cos((M-1)\omega_{in}) & -\sin((M-1)\omega_{in}) & 1 \end{bmatrix} \tag{3.12}$$

The columns of $\mathbf{E}$ are linearly independent and hence $\mathbf{E}$ is a full rank matrix and hence $\mathbf{E}^T\mathbf{E}$ is positive definite [1]. Hence, the second term in (3.10) is nonnegative. Because the first term is positive and does not depend on xp, χ is minimized by choosing $xp = \hat{x}p$. This is equivalent to solve the 3×3 linear system

$$\mathbf{A}\hat{x}p = \mathbf{b} \tag{3.13}$$

with $\mathbf{A} = \mathbf{E}^T\mathbf{E}$ and $\mathbf{b} = \mathbf{E}^T\mathbf{y}$, leading to :

$$\mathbf{A} = \sum_{n=0}^{M-1} \begin{bmatrix} \cos^2(n\omega_{in}) & -0.5\sin(2n\omega_{in}) & \cos(n\omega_{in}) \\ -0.5\sin(2n\omega_{in}) & \sin^2(n\omega_{in}) & -\sin(n\omega_{in}) \\ \cos(n\omega_{in}) & -\sin(n\omega_{in}) & 1 \end{bmatrix} \tag{3.14}$$

---

[1] A matrix $\mathbf{A}$ is positive definite if $\mathbf{z}^T\mathbf{A}\mathbf{z} > 0$ whatever the column vector $\mathbf{z}$. In our case : $\mathbf{z}^T\mathbf{A}\mathbf{z} = (\mathbf{E}\mathbf{z})^T(\mathbf{E}\mathbf{z}) = \|\mathbf{E}\mathbf{z}\|^2 > 0$.

$$b = \sum_{n=0}^{M-1} \begin{bmatrix} \mathbf{y}[n]\cos(n\omega_{in}) \\ -\mathbf{y}[n]\sin(n\omega_{in}) \\ \mathbf{y}[n] \end{bmatrix} \tag{3.15}$$

The matrix $\mathbf{A}$ is symmetric and non-singular and its elements have the following analytical expressions

$$\begin{cases} a_{11} &= \dfrac{M}{2} + \dfrac{\cos((M-1)\omega_{in})\sin(M\omega_{in})}{2\sin(\omega_{in})} \\[2mm] a_{12} &= -\dfrac{\sin((M-1)\omega_{in})\sin(M\omega_{in})}{2\sin(\omega_{in})} \\[2mm] a_{13} &= \dfrac{\cos(\frac{M-1}{2}\omega_{in})\sin(\frac{M}{2}\omega_{in})}{\sin(\frac{\omega_{in}}{2})} \\[2mm] a_{22} &= \dfrac{M}{2} - a_{11} \\[2mm] a_{23} &= -\dfrac{\sin(\frac{M-1}{2}\omega_{in})\sin(\frac{M}{2}\omega_{in})}{\sin(\frac{\omega_{in}}{2})} \\[2mm] a_{33} &= M \end{cases} \tag{3.16}$$

This will reduce the round-off errors and the computation time.

The solution of the linear system (3.13) is given by

$$\begin{cases} \hat{\mathbf{x}}p[1] &= \dfrac{b_1[M(M-a_{11})-a_{23}^2]-b_2[Ma_{12}-a_{13}a_{23}]+b_3[a_{12}a_{23}-a_{13}(M-a_{11})]}{\Delta} \\[3mm] \hat{\mathbf{x}}p[2] &= \dfrac{-b_1[Ma_{12}-a_{12}a_{23}]+b_2[Ma_{11}-a_{13}^2]-b_3[a_{11}a_{23}-a_{12}a_{13}]}{\Delta} \\[3mm] \hat{\mathbf{x}}p[3] &= \dfrac{b_1[a_{12}a_{23}-a_{13}(M-a_{11})]-b_2[a_{11}a_{23}-a_{12}a_{13}]+b_3[a_{11}(M-a_{11})-a_{12}^2]}{\Delta} \end{cases} \tag{3.17}$$

where b_j are the elements of $\mathbf{b}$ and Δ is the determinant of $\mathbf{A}$

$$\Delta = a_{11}[M(M-a_{11})-a_{23}^2] - a_{12}[Ma_{12}-a_{13}a_{23}] \\ + a_{13}[a_{12}a_{23}-a_{13}(M-a_{11})] \tag{3.18}$$

This algorithm could be simplified tacking into account the coherent sampling properties. Indeed, coherent sampling occurs when the sampling and test frequencies satisfy the relationship

$$f_i = \frac{J}{M}f_s \tag{3.19}$$

where J is an integer relatively prime to M. This formula means that there is an integer number of the input sinewave cycles within the data record. If M is a power of two, then it is sufficient to set J equal to an odd integer.

In the case of coherent sampling, the matrix (3.14) is diagonal

$$\mathbf{A} = \frac{M}{2} \begin{bmatrix} 1 & 0 & 0 \\ 0 & 1 & 0 \\ 0 & 0 & 2 \end{bmatrix} \tag{3.20}$$

and the solution of (3.13) becomes

$$\hat{x}p = \frac{1}{M} \sum_{n=0}^{M-1} \begin{bmatrix} 2\mathbf{y}[n]\cos(n\omega_{in}) \\ -2\mathbf{y}[n]\sin(n\omega_{in}) \\ \mathbf{y}[n] \end{bmatrix} \tag{3.21}$$

5. The four-parameter method

When ω_{in} is unknown, then we have to determine the four parameters (3.7). This is done by minimising (3.6) using the Gauss-Newton method [44]

$$\mathbf{x}p_n = \mathbf{x}p_c - \mathbf{F}^{-1}(\mathbf{x}p_c)\mathbf{g}(\mathbf{x}p_c) \tag{3.22}$$

where $\mathbf{x}p_c$ and $\mathbf{x}p_n$ are the current and next estimates of $\mathbf{x}p$; $\mathbf{g}$ and $\mathbf{F}$ are the gradient and the Gauss-Newton approximation of the hessian of the function (3.6). The calculation gives

$$\mathbf{g} = \sum_{n=0}^{M-1} \begin{bmatrix} (\mathbf{x}[n] - \mathbf{y}[n])\cos(n\omega_{in}) \\ -(\mathbf{x}[n] - \mathbf{y}[n])\sin(n\omega_{in}) \\ \mathbf{x}[n] - \mathbf{y}[n] \\ -n(\mathbf{x}[n] - \mathbf{y}[n])v_n \end{bmatrix} \tag{3.23}$$

$$\mathbf{F} = \sum_{n=0}^{M-1} \begin{bmatrix} \cos^2(n\omega_{in}) & -0.5\sin(2n\omega_{in}) \\ -0.5\sin(2n\omega_{in}) & \sin^2(n\omega_{in}) \\ \cos(n\omega_{in}) & -\sin(n\omega_{in}) \\ -nv_n\cos(n\omega_{in}) & nv_n\sin(n\omega_{in}) \end{bmatrix}$$

$$\tag{3.24}$$

$$\begin{bmatrix} \cos(n\omega_{in}) & -nv_n\cos(n\omega_{in}) \\ -\sin(n\omega_{in}) & nv_n\sin(n\omega_{in}) \\ 1 & -nv_n \\ -nv_n & n^2v_n^2 \end{bmatrix}$$

where

$$\mathbf{x}[n] = \mathbf{x}p[1]\cos(n\omega_{in}) - \mathbf{x}p[2]\sin(n\omega_{in}) + \mathbf{x}p[3] \tag{3.25}$$

and

$$v_n = \mathbf{x}p[1]\sin(n\omega_{in}) + \mathbf{x}p[2]\cos(n\omega_{in}) \tag{3.26}$$

The matrix $\mathbf{F}$ is symmetric positive definite and hence non-singular. Its elements have the following expressions

$$
\begin{cases}
f_{11} &= \frac{M}{2} + \frac{1}{2}S_{c,0}(2\omega_{in}) \\
f_{12} &= -\frac{1}{2}S_{s,0}(2\omega_{in}) \\
f_{13} &= S_{c,0}(\omega_{in}) \\
f_{14} &= -\frac{1}{4}\mathbf{xp}[2]M(M-1) - \frac{\mathbf{xp}[1]}{2}S_{s,1}(2\omega_{in}) - \frac{\mathbf{xp}[2]}{2}S_{c,1}(2\omega_{in}) \\
f_{22} &= \frac{M}{2} - f_{11} \\
f_{23} &= -S_{c,0}(\omega_{in}) \\
f_{24} &= \frac{1}{4}\mathbf{xp}[1]M(M-1) + \frac{\mathbf{xp}[2]}{2}S_{s,1}(2\omega_{in}) - \frac{\mathbf{xp}[1]}{2}S_{c,1}(2\omega_{in}) \\
f_{33} &= M \\
f_{34} &= -\mathbf{xp}[1]S_{s,1}(\omega_{in}) - \mathbf{xp}[2]S_{c,1}(\omega_{in}) \\
f_{44} &= \frac{\mathbf{xp}[1]^2 + \mathbf{xp}[2]^2}{12}M(M-1)(2M-1) \\
&+ \frac{\mathbf{xp}[2]^2 - \mathbf{xp}[1]^2}{2}S_{c,2}(2\omega_{in}) + \mathbf{xp}[1]\mathbf{xp}[2]S_{s,2}(2\omega_{in})
\end{cases}
$$

$$(3.27)$$

where

$$
S_{c,m}(\theta) = \sum_{n=0}^{M-1} n^m \cos(n\theta)
$$

$$
S_{s,m}(\theta) = \sum_{n=0}^{M-1} n^m \sin(n\theta)
$$

$$(3.28)$$

$$
m = 0, 1, 2. \ \theta = \omega_{in}, \ 2\omega_{in}
$$

Useful relationships that avoid the summation of (3.28) are

$$
m = 0 \begin{cases}
S_{c,0}(\theta) &= \frac{\cos(\frac{M-1}{2}\theta)\sin(\frac{M}{2}\theta)}{\sin(\frac{\theta}{2})} \\[2ex]
S_{s,0}(\theta) &= \frac{\sin(\frac{M-1}{2}\theta)\sin(\frac{M}{2}\theta)}{\sin(\frac{\theta}{2})}
\end{cases}
$$

$$
m = 1 \begin{cases}
S_{c,1}(\theta) &= \frac{M\sin([M-0.5]\theta) - S_{s,0}(\theta)\cos(\frac{\theta}{2})}{2\sin(\frac{\theta}{2})} - \frac{S_{c,0}(\theta)}{2} \\[2ex]
S_{s,1}(\theta) &= \frac{-M\cos([M-0.5]\theta) + S_{c,0}(\theta)\cos(\frac{\theta}{2})}{2\sin(\frac{\theta}{2})} - \frac{S_{s,0}(\theta)}{2}
\end{cases}
$$

$$(3.29)$$

$$
m = 2 \begin{cases}
S_{c,2}(\theta) &= \frac{M(M-1)\sin([M-0.5]\theta) - 2S_{s,1}(\theta)\cos(\frac{\theta}{2})}{2\sin(\frac{\theta}{2})} \\[2ex]
S_{s,2}(\theta) &= \frac{-M(M-1)\cos([M-0.5]\theta) + 2S_{c,1}(\theta)\cos(\frac{\theta}{2})}{2\sin(\frac{\theta}{2})}
\end{cases}
$$

As in A/D converter testing the residual $(\mathbf{y} - \mathbf{x})$ is small, the Gauss-Newton method is fast linearly locally convergent [44]. Thus, the choice of good starting values $\mathbf{xp}_0$ will guarantee the convergence to the global minimum.

5.1 Influence of the initial frequency

Here we intend to find by means of computer simulations an estimate of the initial frequency accuracy required by the four-parameter method . Then, we will show that the interpolated FFT frequency estimation method guarantees the global convergence of the four-parameter method.

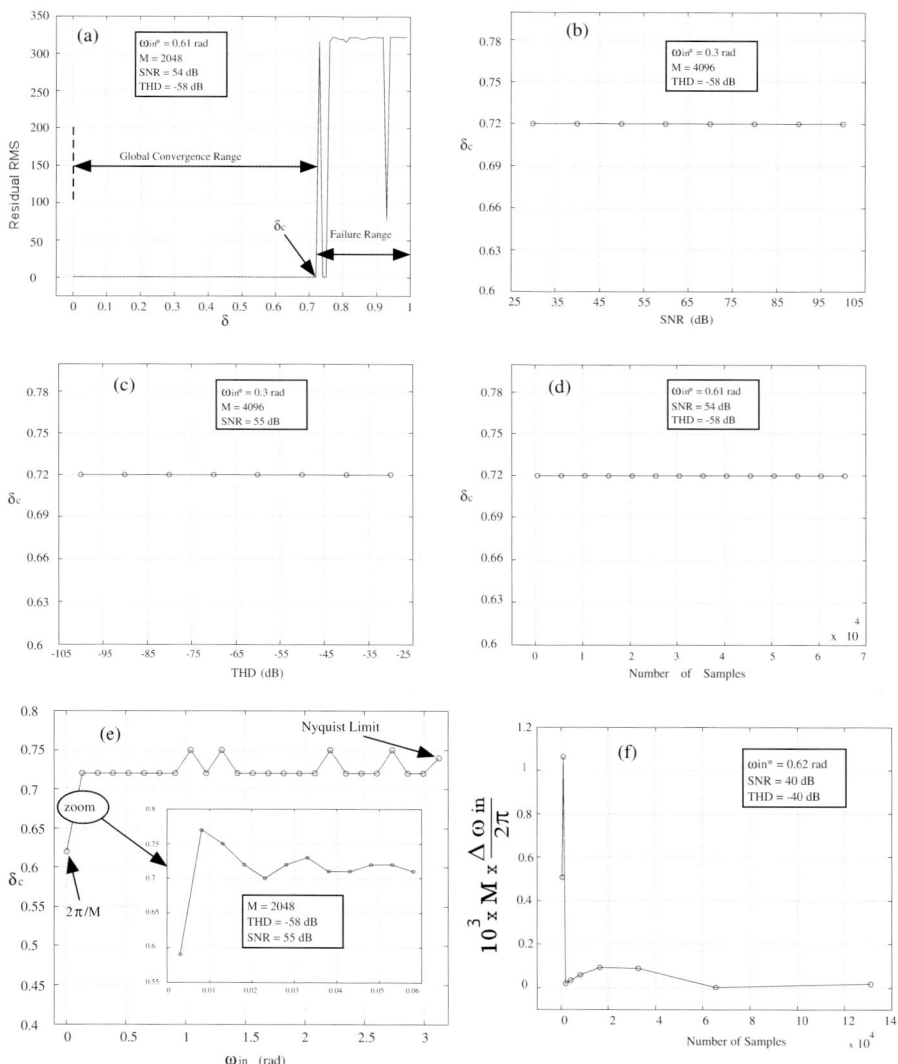

Figure 3.1. (a)-The four-parameter residual RMS versus the initial frequency uncertainty. (b, c, d, e)-Effect of SNR, THD, M, and ω_{in} on δ_c. (f)-The accuracy of the initial frequency estimation method as a function of the number of samples.

Figure 3.1.a shows the variation of the estimated residual RMS

$$\frac{1}{M} \sum_{n=0}^{M-1} (\mathbf{y}[n] - \mathbf{x}[n])^2 \tag{3.30}$$

as a function of the initial frequency uncertainty

$$\delta = \frac{M}{2\pi} |\omega_{in0} - \omega_{in}^\star| \tag{3.31}$$

A critical value δ_c appears and two intervals can be distinguished:

- The convergence range $[0, \delta_c]$ where the algorithm converges to the global minimum;

- The failure range $\delta > \delta_c$ where the algorithm diverges or converges to a local optimum. In this range, we have sometimes a global convergence but in random way.

The variation of δ_c as a function of SNR, THD, M and ω_{in} has been investigated. The obtained results are shown in Figures 3.1.(b, c, d, e). As it can be seen, these parameters have practically no effect on δ_c and 0.7 is a reasonable worst case value. From (3.31), the required initial frequency estimate accuracy is

$$\Delta\omega_{in0} < 2\pi \frac{0.7}{M} \tag{3.32}$$

Figure 3.1.f shows the variation of the estimation error $\Delta\omega_{in} = |\omega_{in} - \omega_{in}^\star|$ of the interpolated FFT and windowing based method versus the number of samples. Worst-case values have been chosen for SNR and THD. It can be seen that for $M \geq 512$

$$\Delta\omega_{in} < 2\pi \frac{0.0012}{M} \tag{3.33}$$

Hence, the criterion (3.32) is largely satisfied and the four-parameter method will converge to the global minimum.

5.2   Estimation of the initial values

The iterative process (3.22) requires good initial values $\mathbf{x}p_0$ which can be estimated as follows:

1 Provide an initial value for ω_{in};

2 Compute the remaining initial values from (3.17).

For this methodology, the Gauss-Newton method requires an initial frequency accuracy of about

$$\frac{\Delta\omega_{in0}}{2\pi} \leq \frac{0.7}{M} \tag{3.34}$$

in order to converge to the global minimum.

In order to satisfy the frequency accuracy (3.34), ω_{in0} can be estimated by the following interpolated FFT and windowing method [68]:

1 Multiply the data $\mathbf{y}$ by the Hanning window

$$\mathbf{w}[n] = 0.5[1 - \cos(2\pi\frac{n}{M})], \quad n = 0, \ldots, M - 1 \tag{3.35}$$

2 Compute the amplitude spectrum $\mathbf{y_{wf}}$ of the windowed data $\mathbf{y}[n]\mathbf{w}[n]$;

3 Find the index J corresponding to the peak of the fundamental component, then evaluate

$$\delta = \frac{2\max(\mathbf{y_{wf}}[J + 1], \mathbf{y_{wf}}[J - 1]) - \mathbf{y_{wf}}[J]}{\mathbf{y_{wf}}[J] + \max(\mathbf{y_{wf}}[J + 1], \mathbf{y_{wf}}[J - 1])} \tag{3.36}$$

4 Finally, the initial frequency value is given by

$$\omega_{in0} = \frac{2\pi}{M}(J + \delta) \tag{3.37}$$

5.3   The algorithm

The Gauss-Newton based method described above can be implemented as follows.

```
Input:

ε the tolerance at which the distance between two successive
    iterates is considered close enough to zero to stop the
    algorithm.  Default value:  [machine accuracy]^(2/3).

k_max maximum number of iterations.  Default value:  30.

Output:

xp estimated parameters

k_f number of iterations


    1 Estimation of the initial values xp_0
```

(a) Compute ω_{in0} from (3.37);

(b) Compute the remaining initial values from (3.17);

2 Gauss-Newton algorithm
   Initialisation: set $\mathbf{x}p_c = \mathbf{x}p_0$;
   FOR $k = 1 : k_{max}$

(a) Compute $\mathbf{g}(\mathbf{x}p_c)$ and $\mathbf{F}(\mathbf{x}p_c)$ from (3.23) and (3.27);

(b) Compute the next iterate $\mathbf{x}p_n$ from (3.22);

(c) If $\max\left\{\left|\frac{\mathbf{x}p_n[i] - \mathbf{x}p_c[i]}{\mathbf{x}p_n[i]}\right|\right\}_{1 \le i \le 4} < \epsilon$, return from here;

(d) Update $\mathbf{x}p_c$: $\mathbf{x}p_c = \mathbf{x}p_n$;

END

$\mathbf{x}p = \mathbf{x}p_n$; $k_f = k$.

6.   Definitions of THD and SNR

The signal-to-noise ratio (SNR) is equal to the ratio of the test signal power to the noise power

$$SNR\,[dB] = 10 \log_{10} \frac{\frac{A^2}{2}}{\frac{1}{M}\sum_{n=0}^{M-1}(\mathbf{y}[n] - \mathbf{x}[n] - \mathbf{h}[n])^2} \qquad (3.38)$$

where $\mathbf{y}$ are the data, $\mathbf{x}$ and $\mathbf{h}$ represent the input sinewave and the harmonics due to the ADC nonlinearity

$$\mathbf{x}[n] = A\cos(n\omega_{in} + \phi) + C \qquad (3.39)$$

$$\mathbf{h}[n] = \sum_{h=2}^{P} A_h \cos(nh\omega_{in} + \phi_h), \; n = 0, \; \ldots, \; M-1 \qquad (3.40)$$

The total harmonic distortion (THD) is equal to the ratio of the harmonic power to the test signal power

$$THD\,[dB] = 10 \log_{10} \frac{\sum_{h=2}^{P} A_h^2}{A^2} \qquad (3.41)$$

7.   The multi-harmonic sine-wave fitting method

The basic idea consists of reducing the problem complexity by separating the linear and nonlinear parameters of the model . The measurement of SNR and THD requires the knowledge of the test signal and harmonic parameters which are: the nonlinear parameter ω_{in} and the linear parameters

$$\mathbf{x}p = [A\cos(\phi),\ A_2\cos(\phi_2), \ldots,\ A_P\cos(\phi_P),$$
$$A\sin(\phi),\ A_2\sin(\phi_2), \ldots,\ A_P\sin(\phi_P),\ C]^T \tag{3.42}$$

The estimates $\mathbf{x}p$ and ω_{in} are obtained by minimising the least-squares cost function

$$\chi(\mathbf{x}p, \omega_{in}) = \sum_{n=0}^{M-1} (\mathbf{y}[n] - \mathbf{x}[n] - \mathbf{h}[n])^2 \tag{3.43}$$

by a variable-separation method [65]: firstly, one estimates the nonlinear parameter ω_{in}. Then, the linear parameters $\mathbf{x}p$ are calculated by solving a linear set of equations.

7.1   Frequency estimation

The data $\mathbf{y}$ can be modelled by the sum of the input sine wave, harmonics and additive noise. As the THD of A/D converters is small, its effect on the frequency accuracy is negligible. Thus, for frequency estimation , one can only consider the input sine wave

$$\mathbf{s}[n] = \mathbf{x}p[1]\cos(n\omega_{in}) - \mathbf{x}p[2]\cos(n\omega_{in}) + \mathbf{x}p[3] \tag{3.44}$$

where

$$\mathbf{x}p = [A\cos(\phi),\ A\sin(\phi),\ C]^T \tag{3.45}$$

For any fixed ω_{in}, χ is minimised by (see paragraph 3.4)

$$\hat{\mathbf{x}}p(\omega_{in}) = \mathbf{A}^{-1}(\omega_{in})\mathbf{b}(\omega_{in}) \tag{3.46}$$

Analytical expressions for the elements of $\mathbf{A}$ are given in section 3.8. Finally, ω_{in} is obtained by minimising the one-variable function

$$L(\omega_{in}) = \chi(\hat{\mathbf{x}}p(\omega_{in}), \omega_{in}) = \mathbf{y}^T\mathbf{y} - \mathbf{b}^T(\omega_{in})\mathbf{A}^{-1}(\omega_{in})\mathbf{b}(\omega_{in}) \tag{3.47}$$

Figure 3.2 shows the variation of $L(\omega_{in})$ around the global minimum $\omega_{in}^\star$. In order to converge to this global minimum, the iterative method used for minimising $L(\omega_{in})$ requires that the initial value ω_{in0} must be in the main lobe, otherwise the routine fails. The main lobe width depends mainly on the

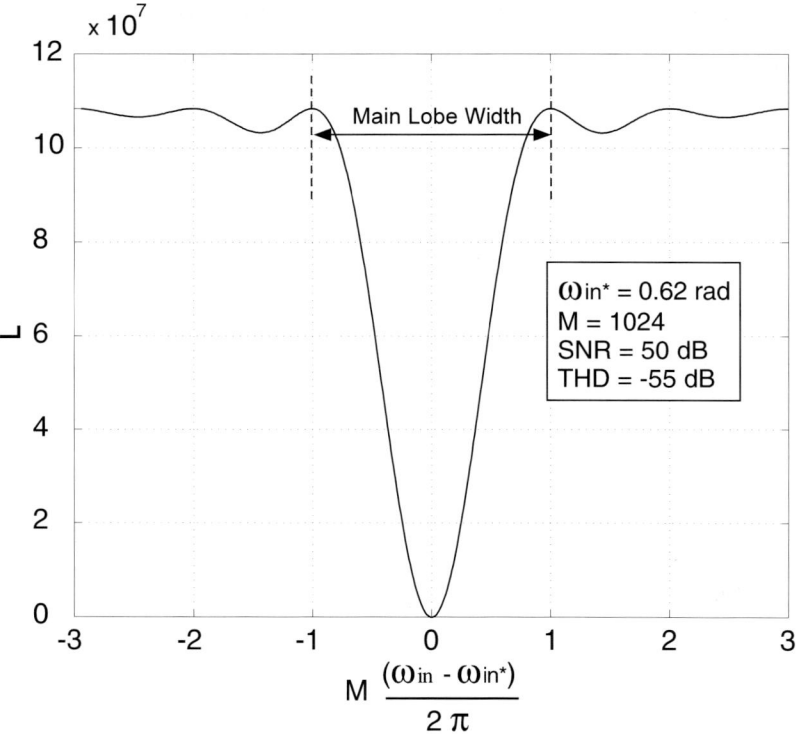

Figure 3.2.   A plot of $L(\omega_{in})$ near the global minimum.

data length M and is equal to $2M$. The necessary condition over ω_{in0} can be expressed by

$$|\omega_{in}^{\star} - \omega_{in0}| < 2\pi \frac{1}{M} \qquad (3.48)$$

ω_{in0} being in the main lobe, the sufficient condition for global convergence is that the routine step $(\omega_{in}^{(k+1)} - \omega_{in}^{k})$ must be in a descent direction

$$(\omega_{in}^{(k+1)} - \omega_{in}^{k})L'(\omega_{in}^{k}) < 0 \qquad (3.49)$$

For the modified Newton method (3.55) given in section 3.8, this condition becomes

$$L'(\omega_{in}^{k}) > 0 \qquad (3.50)$$

Simulations show that this condition is satisfied when the initial frequency uncertainty $|\omega_{in}^{*} - \omega_{in0}|$ is less than $2\pi \frac{0.3}{M}$.

8. Estimation of the normalised angular frequency

The normalised angular frequency ω_{in} is estimated by minimising the one-variable function

$$L(\omega_{in}) = \mathbf{y}^T\mathbf{y} - \mathbf{b}^T(\omega_{in})\mathbf{A}^{-1}(\omega_{in})\mathbf{b}(\omega_{in}) \tag{3.51}$$

where

$$\mathbf{b}(\omega_{in}) = \sum_{n=0}^{M-1} \mathbf{y}[n][\cos(n\omega), -\sin(n\omega), 1]^T \tag{3.52}$$

$$\mathbf{A}(\omega_{in}) = \begin{bmatrix} \frac{M}{2} + \frac{1}{2}S_{c,0}(2\omega_{in}) & -\frac{1}{2}S_{s,0}(2\omega_{in}) & S_{c,0}(\omega_{in}) \\ \star & \frac{M}{2} - \frac{1}{2}S_{c,0}(2\omega_{in}) & -S_{s,0}(\omega_{in}) \\ \star & \star & M \end{bmatrix} \tag{3.53}$$

Matrix $\mathbf{A}$ is symmetric positive definite and hence non-singular. The functions

$$\begin{aligned} S_{c,m}(\theta) &= \sum_{n=0}^{M-1} n^m \cos(n\theta) \\ S_{s,m}(\theta) &= \sum_{n=0}^{M-1} n^m \sin(n\theta) \\ m &= 0,\ 1,\ 2.\ \theta = \omega_{in},\ 2\omega_{in} \end{aligned} \tag{3.54}$$

have analytical expressions given by (3.29).

The minimisation of (3.51) is achieved by the following Newton method

$$\omega_{in}^{(k+1)} = \omega_{in}^k - \frac{L'(\omega_{in}^k)}{L''(\omega_{in}^k)} \tag{3.55}$$

where the derivatives of the function $L(\omega_{in})$ are given by

$$L'(\omega_{in}) = -[\mathbf{b}'^T\mathbf{A}^{-1}\mathbf{b} + \mathbf{b}^T(\mathbf{A}^{-1})'\mathbf{b} + \mathbf{b}^T\mathbf{A}^{-1}\mathbf{b}'] \tag{3.56}$$

$$\begin{aligned} L''(\omega_{in}) = \; & -[\mathbf{b}''^T\mathbf{A}^{-1}\mathbf{b} + \mathbf{b}^T(\mathbf{A}^{-1})''\mathbf{b} + \mathbf{b}^T\mathbf{A}^{-1}\mathbf{b}'' + \\ & 2(\mathbf{b}^T(\mathbf{A}^{-1})'\mathbf{b}' + \mathbf{b}'^T\mathbf{A}^{-1}\mathbf{b}' + \mathbf{b}^T(\mathbf{A}^{-1})'\mathbf{b}')] \end{aligned} \tag{3.57}$$

The derivatives of $\mathbf{A}^{-1}$ as a function of those of $\mathbf{A}$ are obtained by using the fact that $\mathbf{A}\mathbf{A}^{-1}$ is equal to the identity matrix

$$(\mathbf{A}^{-1})' = -\mathbf{A}^{-1}\mathbf{A}'\mathbf{A}^{-1} \tag{3.58}$$

$$(\mathbf{A}^{-1})'' = -[(\mathbf{A}^{-1})'\mathbf{A}'\mathbf{A}^{-1} + \mathbf{A}^{-1}\mathbf{A}''\mathbf{A}^{-1} + \mathbf{A}^{-1}\mathbf{A}'(\mathbf{A}^{-1})'] \tag{3.59}$$

The derivatives of $\mathbf{A}$ and $\mathbf{b}$ have the following expressions

$$\mathbf{b}'(\omega_{in}) = \sum_{n=0}^{M-1} n\mathbf{y}[n][-\sin(n\omega), -\cos(n\omega), 0]^T \tag{3.60}$$

$$\mathbf{b}''(\omega_{in}) = \sum_{n=0}^{M-1} n^2 \mathbf{y}[n][-\cos(n\omega),\ \sin(n\omega),\ 0]^T \qquad (3.61)$$

$$\mathbf{A}' = \begin{bmatrix} -S_{s,1}(2\omega_{in}) & -S_{c,1}(2\omega_{in}) & -S_{s,1}(\omega_{in}) \\ \star & S_{s,1}(2\omega_{in}) & -S_{c,1}(\omega_{in}) \\ \star & \star & 0 \end{bmatrix} \qquad (3.62)$$

$$\mathbf{A}'' = \begin{bmatrix} -2S_{c,2}(2\omega_{in}) & 2S_{s,2}(2\omega_{in}) & -S_{c,2}(\omega_{in}) \\ \star & 2S_{c,2}(2\omega_{in}) & S_{s,2}(\omega_{in}) \\ \star & \star & 0 \end{bmatrix} \qquad (3.63)$$

In order to converge to the global minimum, this frequency estimation method requires an initial frequency accuracy of about

$$\frac{\Delta\omega_{in0}}{2\pi} < \frac{0.3}{M} \qquad (3.64)$$

This condition is largely satisfied by the interpolated FFT method described in 3.5.2.

9.   Estimation of the linear parameters

The estimation of THD and SNR requires the knowledge of the input sinewave and harmonics parameters. In this case, the model to be estimated is

$$\mathbf{s}[n] = \sum_{h=1}^{P} A_h \cos(nh\omega_{in} + \phi_h) + C,\ n = 0,\ \ldots,\ M-1 \qquad (3.65)$$

The normalised angular frequency being determined, the minimisation of the function (3.43) with respect to the linear parameters (3.42) yields

$$\mathbf{E}_P^T(\omega_{in})\mathbf{E}_P(\omega_{in})\hat{x}p = \mathbf{E}_P^T(\omega_{in})\mathbf{y} \qquad (3.66)$$

where $\mathbf{E}_P$ is a M-by-$(2P+1)$ matrix whose elements are given by

$$e_{ij} = \begin{cases} \cos(j[i-1]\omega_{in}) & 1 \le j \le P \\ -\sin([i-1][j-P]\omega_{in}) & P+1 \le j \le 2P \\ 1 & j = 2P+1 \\ i = 1,\ldots,M \end{cases} \qquad (3.67)$$

In practice, the matrix $\mathbf{E}_P^T\mathbf{E}_P$ is practically always nonsingular, and hence the linear system (3.66) can be solved by the Cholesky decomposition technique [120]. The solution of (3.66) should be implemented so as to avoid the storage of the large size matrix $\mathbf{E}_P$

$$(\mathbf{E}_P^T\mathbf{E}_P)[i,j] = \sum_{k=0}^{M-1} e_{ki}e_{kj},\ i,\ j = 1,\ \ldots,\ 2P+1 \qquad (3.68)$$

$$(\mathbf{E}_P^T \mathbf{y})[j] = \begin{cases} \displaystyle\sum_{n=0}^{M-1} y_n \cos(n j \omega_{in}) & 1 \le j \le P \\[2mm] -\displaystyle\sum_{n=0}^{M-1} y_n \sin(n(j-P)\omega_{in}) & P+1 \le j \le 2P \\[2mm] \displaystyle\sum_{n=0}^{M-1} y_n & j = 2P+1 \end{cases} \qquad (3.69)$$

Analytical expressions for the elements of $\mathbf{E}_P^T \mathbf{E}_P$ can be obtained from (3.29). An other solution consists of solving the overdetermined system $\mathbf{E}_P(\omega_{in})\hat{\mathbf{x}}_P = \mathbf{y}$ by the singular value decomposition [120]. This method may be slow, especially when the number of samples is large, but works even when $\mathbf{E}_P$ is rank deficient.

10. On the rank of $\mathbf{E}_P$

Here we show that the matrix $\mathbf{E}_P(\omega_{in})$ has full rank $2P+1$ if ω_{in} leads at least $2P+1$ to distinct complex numbers $z_n = \exp(jn\omega_{in})$, $0 \le n \le M-1$. This situation occurs in particular when $\omega_{in} < \frac{\pi}{P}$ or in the case of coherent sampling: $\omega_{in} = 2\pi\frac{J}{M}$ where J is an integer relatively prime with M.

Suppose that there exist constants λ_k such that $\displaystyle\sum_{k=1}^{2P+1} \lambda_k \mathbf{c}_k = 0$, where $\mathbf{c}_k$ are the columns of $\mathbf{E}_P$. From (3.67), this implies that

$$\sum_{k=1}^{P} [\lambda_k \cos(kn\omega_{in}) - \lambda_{k+P} \sin(kn\omega_{in})] + \lambda_{2P+1} = 0 \qquad (3.70)$$

for all $n = 0, \ldots, M-1$. (3.70) can be rewritten as

$$\sum_{k=0}^{2P} \gamma_k z_n^k = 0, \ 0 \le n \le M-1 \qquad (3.71)$$

with

$$\gamma_k = \begin{cases} \lambda_{(P-k)} - j\lambda_{(2P-k)} & 0 \le k \le P-1 \\ 2\lambda_{(2P+1)} & k = P \\ \lambda_{(k-P)} + j\lambda_k & P+1 \le k \le 2P \end{cases} \qquad (3.72)$$

From (3.71), the polynomial $Q(z) = \displaystyle\sum_{k=0}^{2P} \gamma_k z^k$ has M zeros z_n while its degree is $2P$. It is only possible when $\gamma_k = 0$ for all k, if we assume that there are at least $2P+1$ distinct roots z_n. From (3.72), one obtains that all λ_k are equal to zero. As a result, the columns of $\mathbf{E}_P$ are linear independent and hence its rank is equal to $2P+1$.

If $\omega_{in} < \frac{\pi}{P}$, then $n\omega_{in} < \pi$ for all $0 \leq n \leq P$ and hence the corresponding $P + 1$ roots z_n are distinct. The polynomial $Q(z)$ is complex, then it has also the following distinct roots $z_n^* = z_{(-n)}$, $1 \leq n \leq P$. Thus, we have $2P + 1$ distinct roots and hence $\mathbf{E}_P$ has full rank for any $\omega_{in} < \frac{\pi}{P}$.

Suppose that there exist $n_1 > n_2$ such that their corresponding roots are equal: $\exp(jn_1\omega_{in}) = \exp(jn_2\omega_{in})$. This implies that $\omega_{in} = 2\pi\frac{k}{n_1 - n_2}$ with $k = 1, \ldots, (n_1 - n_2)$ (because $0 < \omega_{in} \leq \pi$). It is only possible when ω_{in} does not satisfy the coherent sampling condition. This equation also gives the form of the values of ω_{in} which result in a rank deficient matrix.

11.   The algorithm

The parameter estimation method described above can be implemented as follows .

Input Parameters:

ϵ the tolerance at which the distance between two successive
    iterates is considered close enough to zero to stop the
    algorithm. Default value: $[\text{machine accuracy}]^{\frac{2}{3}}$.

k_{max} maximum number of iteration. Default value: 30.

Output parameters:

ω_{in} estimated normalised angular frequency

$\mathbf{x}p$ estimated linear parameters

k_f number of iterations

   1 Determine the initial frequency ω_{in0} as explained in
     section 3.5;

   2 Frequency estimation:
     Set $\omega_{in}^c = \omega_{in0}$;
     FOR $k = 1 : k_{max}$

     (a) Compute $\mathbf{A}(\omega_{in}^c)$, $\mathbf{A}'(\omega_{in}^c)$, $\mathbf{A}''(\omega_{in}^c)$, $\mathbf{b}(\omega_{in}^c)$, $\mathbf{b}'(\omega_{in}^c)$, $\mathbf{b}''(\omega_{in}^c)$,
         from (3.53), (3.62), (3.63), (3.52), (3.60), (3.61);
     (b) Compute $(\mathbf{A}^{-1})'(\omega_{in}^c)$ and $(\mathbf{A}^{-1})''(\omega_{in}^c)$ from (3.58), (3.59);
     (c) Calculate $L'(\omega_{in}^c)$ and $L''(\omega_{in}^c)$ from (3.56), (3.57);
     (d) Calculate the next iterate ω_{in}^n from (3.55);
     (e) If $|\omega_{in}^n - \omega_{in}^c| < \epsilon\omega_{in}^n$, return from here;
     (f) Update ω_{in}^c: $\omega_{in}^c = \omega_{in}^n$;

```
END
```
$$\omega_{in} = \omega_{in}^n, \quad k_f = k;$$

3 Estimate the linear parameters xp from (3.66);

12. Multitone test to circumvent signal purity problems

The A/D converter test in single-tone mode requires a spectrally pure input signal test . This condition is no longer fulfilled for high resolution converters. The solution which consists of using an analogue filter can be a costly hard-to-apply method. An other alternative based on the use of a dual-tone test signal is proposed [19].

12.1 Test method

The setup of this dual-tone procedure is similar to that of the intermodulation distortion measurement. Assume that we want to characterize the converter in single-tone mode with a sine wave whose amplitude and frequency are respectively equal to $2A$ and f_0. The equivalent dual-tone test procedure consists of stimulating the converter by a sum of two sine waves with equal amplitudes but with different frequencies f_1 and f_2. The amplitude of each sine wave equals A and the frequencies f_1 and f_2 are close to f_0 and chosen so as to satisfy the coherent sampling condition.

12.2 Relation between single-tone and dual-tone parameters

From the ADC output data compute the intermodulation distortion IMD. For signal generators whose harmonic distortion is less -30 dBc, which is usually the case, IMD is an intrinsic characteristic of the converter under test, since the effect of the the generator distortion is negligible. The total harmonic distortion in single-tone mode is equal to the intermodulation distortion thus obtained with an approximation of about 0.5 dB. Also the effective number of bits in single-tone mode is equal to that in dual-tone mode with an error less than 0.5 bits. The effective number of bits (N_{ef}^{DT}) and the signal-to-noise and distortion ratio $(SINAD_{DT})$ in dual-tone mode are related by the following relationship

$$N_{ef}^{DT} = \frac{SINAD_{DT} + 1.25}{6.02} \qquad (3.73)$$

$SINAD_{DT}$ is equal to the ratio of the dual-tone test signal power to the noise and distortion (including intermodulations) power expressed in dB. In the spectral domain $SINAD_{DT}$ is given by

$$SINAD_{DT} \, [dB] = 10 \log \left(\frac{P(f_1) + P(f_2)}{P_{nd}} \right) \qquad (3.74)$$

where $P(f_1$ and $P(f_2)$ are the spectral powers of the fundamental components; P_{nd} is the spectral power of the noise and distortion obtained by setting the spectral lines corresponding to the fundamental components and DC offset equal to zero.

When the test signal is not full scale, i.e. A is less than a quarter of the full-scale range (FSR), $SINAD_{DT}$ value must be corrected by subtracting from it $10\log(\frac{4A}{FSR})$. For the issues concerning the leakage effect and windowing refer to the Frequency-Domain Data Analysis chapter.

12.3   Parameter calculation in the time domain

The estimation of the ADC parameters by the spectral analysis suffers from the leakage effect. In order to overcome this problem, we propose here a time-domain algorithm for estimating the ADC parameters in dual-mode. We consider the general case where the amplitudes of the input sine waves are different. In the time domain the estimation of the ADC parameters amounts to the identification of harmonic and intermodulation tones in noise. The use of the variable projection method [65] reduces the identification problem to a 2-dimensional nonlinear least squares (NLS) estimation of the two normalised angular frequencies (NAFs) ω_{in1} and ω_{in2} of the inputsine waves, followed by solving a linear system in order to estimate the amplitudes and phases of the different tones.

The NLS frequency estimation problem complexity can be reduced by neglecting the distortion terms. This approximation is reasonable since the distortion of A/D converters is small and in general the contribution of the noise is dominant compared to that of the distortion. In this case the NLS cost function to be minimised in order to estimate the NAFs is given by

$$L(\mathbf{w}) = \mathbf{y}^T\mathbf{y} - \mathbf{b}^T(\mathbf{w})\mathbf{A}^{-1}(\mathbf{w})\mathbf{b}(\mathbf{w}) \tag{3.75}$$

with $\mathbf{w} = [\omega_{in1}, \omega_{in2}]^T$, $\mathbf{b}(\mathbf{w}) = \mathbf{E}^T(\mathbf{w})\mathbf{y}$, $\mathbf{A}(\mathbf{w}) = \mathbf{E}^T(\mathbf{w})\mathbf{E}(\mathbf{w})$ where T denotes the transpose operator; $\mathbf{y}$ is a $M \times 1$ column vector whose components are the data samples, M being the number of samples.

E is the following $M \times 5$ matrix

$$
\mathbf{E(w)} =
\begin{pmatrix}
1 & 0 \\
\cos(\omega_{in1}) & \sin(\omega_{in1}) \\
\cos(2\omega_{in1}) & \sin(2\omega_{in1}) \\
\vdots & \vdots \\
\cos((M-1)\omega_{in1}) & \sin((M-1)\omega_{in1})
\end{pmatrix}
$$

$$
\begin{pmatrix}
1 & 0 & 1 \\
\cos(\omega_{in2}) & \sin(\omega_{in2}) & 1 \\
\cos(2\omega_{in2}) & \sin(2\omega_{in2}) & 1 \\
\vdots & \vdots & \vdots \\
\cos((M-1)\omega_{in2}) & \sin((M-1)\omega_{in2}) & 1
\end{pmatrix}
$$

(3.76)

The matrix $\mathbf{E}$ is full rank and hence $\mathbf{A}$ is nonsingular.

12.3.1   Frequency estimation algorithm.   The use of the Newton method to minimise the cost function $L(\mathbf{w})$ leads to the following iterative scheme [44]

$$
\begin{cases}
\omega_{in1}^{k+1} = \omega_{in1}^{k} + \Delta\omega_{in1}^{k} \\[2mm]
\omega_{in2}^{k+1} = \omega_{in2}^{k} + \Delta\omega_{in2}^{k}
\end{cases}
$$

(3.77)

where the frequency steps $\Delta\omega_{in1}^{k}$ and $\Delta\omega_{in2}^{k}$ are given by

$$
\begin{cases}
\Delta\omega_{in1}^{k} = \dfrac{\left(\dfrac{\partial^2 L}{\partial\omega_{in1}\partial\omega_{in2}}\dfrac{\partial L}{\partial\omega_{in2}}\right) - \left(\dfrac{\partial^2 L}{\partial\omega_{in2}^2}\dfrac{\partial L}{\partial\omega_{in1}}\right)}{\left(\dfrac{\partial^2 L}{\partial\omega_{in1}^2}\dfrac{\partial^2 L}{\partial\omega_{in2}^2}\right) - \left(\dfrac{\partial^2 L}{\partial\omega_{in1}\partial\omega_{in2}}\right)^2} \\[6mm]
\Delta\omega_{in2}^{k} = \dfrac{\left(\dfrac{\partial^2 L}{\partial\omega_{in1}\partial\omega_{in2}}\dfrac{\partial L}{\partial\omega_{in1}}\right) - \left(\dfrac{\partial^2 L}{\partial\omega_{in1}^2}\dfrac{\partial L}{\partial\omega_{in2}}\right)}{\left(\dfrac{\partial^2 L}{\partial\omega_{in1}^2}\dfrac{\partial^2 L}{\partial\omega_{in2}^2}\right) - \left(\dfrac{\partial^2 L}{\partial\omega_{in1}\partial\omega_{in2}}\right)^2}
\end{cases}
$$

(3.78)

The partial derivatives of the cost function are evaluated at $\mathbf{w}_k = [\omega_{in1}^k, \omega_{in2}^k]^T$ and given by

$$
\begin{cases}
-\dfrac{\partial L}{\partial \omega_{in,i}} = \left(\dfrac{\partial \mathbf{b}}{\partial \omega_{in,i}}\right)^T \mathbf{A}^{-1}\mathbf{b} + \mathbf{b}^T \dfrac{\partial \mathbf{A}^{-1}}{\partial \omega_{in,i}}\mathbf{b} + \mathbf{b}^T \mathbf{A}^{-1}\dfrac{\partial \mathbf{b}}{\partial \omega_{in,i}} \\[2em]
-\dfrac{\partial^2 L}{\partial \omega_{in,i}^2} = \left(\dfrac{\partial^2 \mathbf{b}}{\partial \omega_{in,i}^2}\right)^T \mathbf{A}^{-1}\mathbf{b} + \mathbf{b}^T \dfrac{\partial^2 \mathbf{A}^{-1}}{\partial \omega_{in,i}^2}\mathbf{b} + \mathbf{b}^T \mathbf{A}^{-1}\dfrac{\partial^2 \mathbf{b}}{\partial \omega_{in,i}^2} + \\[1em]
\qquad\quad 2\left[\left(\dfrac{\partial \mathbf{b}}{\partial \omega_{in,i}}\right)^T \dfrac{\partial \mathbf{A}^{-1}}{\partial \omega_{in,i}}\mathbf{b} + \left(\dfrac{\partial \mathbf{b}}{\partial \omega_{in,i}}\right)^T \mathbf{A}^{-1}\dfrac{\partial \mathbf{b}}{\partial \omega_{in,i}} + \right. \\[1em]
\qquad\quad \left. \mathbf{b}^T \dfrac{\partial \mathbf{A}^{-1}}{\partial \omega_{in,i}}\dfrac{\partial \mathbf{b}}{\partial \omega_{in,i}}\right] \\[2em]
-\dfrac{\partial^2 L}{\partial \omega_{in1}\partial \omega_{in2}} = \mathbf{b}^T \dfrac{\partial^2 \mathbf{A}^{-1}}{\partial \omega_{in1}\partial \omega_{in2}}\mathbf{b} + \left(\dfrac{\partial \mathbf{b}}{\partial \omega_{in1}}\right)^T \dfrac{\partial \mathbf{A}^{-1}}{\partial \omega_{in2}}\mathbf{b} \\[1em]
\qquad\quad + \left(\dfrac{\partial \mathbf{b}}{\partial \omega_{in1}}\right)^T \mathbf{A}^{-1}\dfrac{\partial \mathbf{b}}{\partial \omega_{in2}} + \left(\dfrac{\partial \mathbf{b}}{\partial \omega_{in2}}\right)^T \dfrac{\partial \mathbf{A}^{-1}}{\partial \omega_{in1}}\mathbf{b} \\[1em]
\qquad\quad + \mathbf{b}^T \dfrac{\partial \mathbf{A}^{-1}}{\partial \omega_{in1}}\dfrac{\partial \mathbf{b}}{\partial \omega_{in2}} + \left(\dfrac{\partial \mathbf{b}}{\partial \omega_{in2}}\right)^T \mathbf{A}^{-1}\dfrac{\partial \mathbf{b}}{\partial \omega_{in1}} \\[1em]
\qquad\quad + \mathbf{b}^T \dfrac{\partial \mathbf{A}^{-1}}{\partial \omega_{in2}}\dfrac{\partial \mathbf{b}}{\partial \omega_{in1}}
\end{cases}
$$

$$(3.79)$$

with $i = 1,\ 2$. The vector $\mathbf{b}$ and its partial derivatives have the following expressions

$$
\begin{cases}
\mathbf{b} = \displaystyle\sum_{n=0}^{M-1} y_n[\cos(n\omega_{in1})\ \sin(n\omega_{in1})\ 0\ 0\ 0]^T \\[1.5em]
\dfrac{\partial \mathbf{b}}{\partial \omega_{in1}} = \displaystyle\sum_{n=0}^{M-1} ny_n[-\sin(n\omega_{in1})\ \cos(n\omega_{in1})\ 0\ 0\ 0]^T \\[1.5em]
\dfrac{\partial \mathbf{b}}{\partial \omega_{in2}} = \displaystyle\sum_{n=0}^{M-1} ny_n[0\ 0\ -\sin(n\omega_{in2})\ \cos(n\omega_{in2})\ 0]^T \\[1.5em]
\dfrac{\partial^2 \mathbf{b}}{\partial \omega_{in1}^2} = \displaystyle\sum_{n=0}^{M-1} n^2 y_n[-\cos(n\omega_{in1})\ -\sin(n\omega_{in1})\ 0\ 0\ 0]^T \\[1.5em]
\dfrac{\partial^2 \mathbf{b}}{\partial \omega_{in2}^2} = \displaystyle\sum_{n=0}^{M-1} n^2 y_n[0\ 0\ -\cos(n\omega_{in2})\ -\sin(n\omega_{in2})\ 0]^T
\end{cases}
$$

$$(3.80)$$

The partial derivatives of $\mathbf{A}^{-1}$ as a function of those $\mathbf{A}$ are obtained by using the fact that $\mathbf{A}\mathbf{A}^{-1}$ equals the identity matrix. The derivation of this equality yields

$$
\begin{cases}
\dfrac{\partial \mathbf{A}^{-1}}{\partial w_{in,i}} = -\mathbf{A}^{-1}\dfrac{\partial \mathbf{A}}{\partial w_{in,i}}\mathbf{A}^{-1} \\[4mm]
\dfrac{\partial^2 \mathbf{A}^{-1}}{\partial w_{in,i}^2} = -\left[\dfrac{\partial \mathbf{A}^{-1}}{\partial w_{in,i}}\dfrac{\partial \mathbf{A}}{\partial w_{in,i}}\mathbf{A}^{-1} + \mathbf{A}^{-1}\dfrac{\partial^2 \mathbf{A}}{\partial w_{in,i}^2}\mathbf{A}^{-1}\right. \\[4mm]
\qquad\qquad \left. + \mathbf{A}^{-1}\dfrac{\partial \mathbf{A}}{\partial w_{in,i}}\dfrac{\partial \mathbf{A}^{-1}}{\partial w_{in,i}}\right] \\[4mm]
\dfrac{\partial^2 \mathbf{A}^{-1}}{\partial w_{in1}\partial w_{in1}} = -\left[\dfrac{\partial \mathbf{A}^{-1}}{\partial w_{in2}}\dfrac{\partial \mathbf{A}}{\partial w_{in1}}\mathbf{A}^{-1} + \mathbf{A}^{-1}\dfrac{\partial^2 \mathbf{A}}{\partial w_{in1}\partial w_{in2}}\mathbf{A}^{-1}\right. \\[4mm]
\qquad\qquad \left. + \mathbf{A}^{-1}\dfrac{\partial \mathbf{A}}{\partial w_{in1}}\dfrac{\partial \mathbf{A}^{-1}}{\partial w_{in2}}\right]
\end{cases}
\tag{3.81}
$$

The matrix $\mathbf{A}$ has the following expression

$$
\left[
\begin{array}{cc}
\frac{M}{2}+0.5S_{c,0}(2w_{in1}) & -0.5S_{s,0}(2w_{in1}) \\
\star & \frac{M}{2}-0.5S_{c,0}(2w_{in1}) \\
\star & \star \\
\star & \star \\
\star & \star \\
\end{array}
\right.
$$

$$
\begin{array}{c}
0.5[S_{c,0}(w_{in1}-w_{in2})+S_{c,0}(w_{in1}+w_{in2})] \\
0.5[S_{s,0}(w_{in1}-w_{in2})+S_{s,0}(w_{in1}+w_{in2})] \\
\frac{M}{2}+0.5S_{c,0}(2w_{in2}) \\
\star \\
\star
\end{array}
\tag{3.82}
$$

$$
\left.
\begin{array}{cc}
0.5[S_{s,0}(-w_{in1}+w_{in2})+S_{s,0}(w_{in1}+w_{in2})] & S_{c,0}(w_{in1}) \\
0.5[S_{c,0}(w_{in1}-w_{in2})+S_{c,0}(w_{in1}+w_{in2})] & S_{s,0}(w_{in1}) \\
0.5S_{s,0}(w_{in2}) & S_{c,0}(w_{in2}) \\
\frac{M}{2}-0.5S_{c,0}(2w_{in2}) & S_{s,0}(w_{in2}) \\
\star & M
\end{array}
\right]
$$

The elements denoted by the symbol $\star$ are obtained by using the fact that $\mathbf{A}$ is symmetric; the functions $S_{c,m}$ and $S_{s,m}$, $m = 0,\ 1,\ 2$, are given by (3.29). The partial derivatives of $\mathbf{A}$ are obtained from (3.82) by using the following simple derivation rules

$$
\begin{aligned}
\frac{\partial S_{c,m}(\alpha w_{in,i}+\beta w_{in,j})}{\partial w_{in,i}} &= -\alpha S_{s,m+1}(\alpha w_{in,i}+\beta w_{in,j}) \\
\frac{\partial S_{s,m}(\alpha w_{in,i}+\beta w_{in,j})}{\partial w_{in,i}} &= \alpha S_{c,m+1}(\alpha w_{in,i}+\beta w_{in,j})
\end{aligned}
\tag{3.83}
$$

with $i \neq j = 1,\ 2;\ m = 0,\ 1$

Given initial values ω_{in1}^0 and ω_{in2}^0, the process (3.77) is iterated until the following condition is satisfied

$$\max \left\{ \left| \frac{\Delta \omega_{in1}^k}{\omega_{in1}^k} \right|, \left| \frac{\Delta \omega_{in2}^k}{\omega_{in2}^k} \right| \right\} \leq (\text{machine accuracy})^{\frac{2}{3}} \qquad (3.84)$$

Use the interpolated fast Fourier transform method described in [68] and summarized in 5.2 in order to generate the frequency initial values. The accuracy of this method is sufficient to guarantee the convergence to the global minimum of the frequency estimation algorithm.

Once the frequency estimates $\hat{\omega}_{in1}$ and $\hat{\omega}_{in2}$ are obtained, calculate the power of the noise plus distortion by

$$P_{nd} = \frac{1}{M} L(\hat{\omega}_{in1}, \hat{\omega}_{in2}) \qquad (3.85)$$

12.3.2   Amplitudes and phases of input sine waves.   Let A_1, A_2, ϕ_1 and ϕ_2 be the amplitudes and phases of the test sine waves. Consider the column vector

$$\mathbf{x}_p = [A_1 \cos(\phi_1), \ -A_1 \sin(\phi_1), \ A_2 \cos(\phi_2), \ -A_2 \sin(\phi_2), \ C]^T \qquad (3.86)$$

where C is the DC offset. The least squares (LS) estimate of $\mathbf{x}_p$ is obtained by solving the following linear system

$$\mathbf{A}(\hat{\omega}_{in1}, \hat{\omega}_{in2})\hat{\mathbf{x}}_p = \mathbf{b}(\hat{\omega}_{in1}, \hat{\omega}_{in2}) \qquad (3.87)$$

where $\mathbf{b}$ and $\mathbf{A}$ are given by (3.80) and (3.82). The power of the test signal is given by

$$P_t = \frac{\sum\limits_{k=1}^{4} \hat{\mathbf{x}}_p^2[k]}{2} \qquad (3.88)$$

12.3.3   Calculation of $SINAD_{DT}$ and N_{ef}^{DT}.   By definition, $SINAD_{DT}$ is equal to the ratio of the test signal power to the noise and distortion power

$$SINAD_{DT} \ [dB] = 10 \log \left(\frac{P_t}{P_{nd}} \right) \qquad (3.89)$$

where P_{nd} and P_t are given by (3.85) and (3.88). In the time domain the effective number of bits is defined by

$$N_{ef}^{DT} \ [Bits] = N - \log_2 \sqrt{12 P_{nd}} \qquad (3.90)$$

where N is the ADC number of bits.

The algorithm thus described for estimating $SINAD_{DT}$ and N_{ef}^{DT} is accurate (because the signal to noise ratio is high) and fast even when the number of samples is large, that is, when dealing with medium-high resolution ADCs. The speed of the algorithm is due to

1 All the matrices and vectors used in the algorithm are of small size and independent from the number of samples;

2 The fast convergence of the Newton method;

3 The interpolated FFT method generates good frequency starting values.

12.3.4   Amplitudes and phases of harmonic and intermodulation tones.

The test signal parameters being determined, remove it from the ADC output data $\mathbf{y}$. One obtains new data $\tilde{\mathbf{y}}$ which consists of harmonic and intermodulation tones in noise. For the tone with NAF ω, amplitude A and phase ϕ, we consider the new parameters

$$\alpha(\omega) = A\cos(\phi), \ \beta(\omega) = -A\sin(\phi) \tag{3.91}$$

Let P_1 and P_2 be the largest harmonic orders corresponding to ω_{in1} and ω_{in2}, and let P denote the largest intermodulation term order taken into account. The intermodulation NAFs are given by

$$\begin{cases} \omega_{in,ij}^{+} = i\omega_{in1} + j\omega_{in2}, \ \omega_{in,ij}^{-} = |i\omega_{in1} - j\omega_{in2}| \\[2mm] \text{with } i > 0, \ j > 0, 2 \le i + j \le P \end{cases} \tag{3.92}$$

The linear parameters of harmonic and intermodulation tones are

$$\begin{aligned} \tilde{\mathbf{x}}_p = \ & [\alpha(2\omega_{in1}) \cdots \alpha(P_1\omega_{in1}), \beta(2\omega_{in1}) \cdots \beta(P_1\omega_{in1}), \\ & \alpha(2\omega_{in2}) \cdots \alpha(P_2\omega_{in2}), \beta(2\omega_{in2}) \cdots \beta(P_2\omega_{in2}), \\ & \left\{\alpha(\omega_{in,ij}^{+})\right\}_{i+j=2}, \left\{\beta(\omega_{in,ij}^{+})\right\}_{i+j=2} \left\{\alpha(\omega_{in,ij}^{-})\right\}_{i+j=2}, \\ & \left\{\beta(\omega_{in,ij}^{-})\right\}_{i+j=2}, \left\{\alpha(\omega_{in,ij}^{+})\right\}_{i+j=P}, \\ & \left\{\beta(\omega_{in,ij}^{+})\right\}_{i+j=P} \left\{\alpha(\omega_{in,ij}^{-})\right\}_{i+j=P}, \left\{\beta(\omega_{in,ij}^{-})\right\}_{i+j=P}\Big]^{T} \end{aligned} \tag{3.93}$$

Let $\tilde{\mathbf{E}}(\hat{\mathbf{w}})$ be the matrix defined as follows

$$
\begin{cases}
\tilde{\mathbf{E}} = [\mathbf{B}(\hat{\omega}_{in1}), \mathbf{C}(\hat{\omega}_{in2}), \mathbf{D}(\hat{\mathbf{w}})] \text{, with} \\[2ex]
\mathbf{B} = \left[\{\cos(h\hat{\omega}_{in1}\mathbf{t})\}_{2 \leq h \leq P_1}, \{\sin(h\hat{\omega}_{in1}\mathbf{t})\}_{2 \leq h \leq P_1} \right] \\[2ex]
\mathbf{C} = \left[\{\cos(h\hat{\omega}_{in2}\mathbf{t})\}_{2 \leq h \leq P_2}, \{\sin(h\hat{\omega}_{in2}\mathbf{t})\}_{2 \leq h \leq P_2} \right] \\[2ex]
\mathbf{D} = [\mathbf{D}_2, \mathbf{D}_3, \cdots, \mathbf{D}_P]
\end{cases}
\tag{3.94}
$$

where $\mathbf{t}$ is the column vector $[0, 1, \cdots, M-1]^T$ and

$$
\mathbf{D}_q(\hat{\mathbf{w}}) = \left[\left\{\cos(\omega_{in,ij}^+\mathbf{t})\right\}_{i+j=q}, \left\{\sin(\omega_{in,ij}^+\mathbf{t})\right\}_{i+j=q}, \right.
$$
$$
\left. \left\{\cos(\omega_{in,ij}^-\mathbf{t})\right\}_{i+j=q}, \left\{\sin(\omega_{in,ij}^-\mathbf{t})\right\}_{i+j=q} \right]
\tag{3.95}
$$

with $2 \leq q \leq P$.

The least squares estimate of $\tilde{\mathbf{x}}_p$ is given by

$$
\hat{\tilde{\mathbf{x}}}_p = \tilde{\mathbf{E}}^+(\hat{\mathbf{w}})\tilde{\mathbf{y}}
\tag{3.96}
$$

where $\tilde{\mathbf{E}}^+$ is the pseudo-inverse of $\tilde{\mathbf{E}}$. In order to save execution time, especially when the number of samples is large, calculate $\tilde{\mathbf{E}}^+$ as follows. If $\tilde{\mathbf{E}}^T\tilde{\mathbf{E}}$ is well-conditioned $\tilde{\mathbf{E}}^+ = [\tilde{\mathbf{E}}^T\tilde{\mathbf{E}}]^{-1}\tilde{\mathbf{E}}^T$, otherwise compute $\tilde{\mathbf{E}}^+$ by the singular value decomposition (SVD) [120].

The power of the intermodulation tones is given by

$$
P_{im} = \frac{1}{2} \sum_{k \geq 2P_1 + 2P_2 - 3} \hat{\tilde{x}}_p^2[k]
\tag{3.97}
$$

12.3.5   Calculation of IMD.   The intermodulation distortion is defined as the ratio of the power of the intermodulation tones to that of the dual-tone test signal

$$
IMD\,[dB] = 10\log\left(\frac{P_{im}}{P_t}\right)
\tag{3.98}
$$

where P_t and P_{im} are given by (3.88) and (3.97).

Chapter 4

FREQUENCY-DOMAIN DATA ANALYSIS

Pierre-Yves Roy

now with
EADS Defence and Security Systems SA
Defence and Communications Systems
Rue Jean-Pierre Timbaud - Montigny le Bretonneux
78063 Saint Quentin Yvelines Cedex, France

pierre-yves.roy@eads-telecom.com

Jacques Durand

now retired from
THALES
L'Orée de Corbeville, BP 56 91401 Orsay, France

1. Discrete Fourier Transform and Fast Fourier Transform

The Discrete Fourier Transform (DFT) of a record of data $y[n]$ that is M samples long, is defined by

$$Y[k] = \sum_{n=0}^{M-1} y[n]\, e^{-i2\pi n \frac{k}{M}} \qquad k = 0 \ldots M - 1 \qquad (4.1)$$

The reverse DFT is

$$y[n] = \frac{1}{M} \sum_{k=0}^{M-1} Y[k]\, e^{+i2\pi k \frac{n}{M}} \qquad n = 0 \ldots M - 1 \qquad (4.2)$$

The evaluation of an ADC in the frequency domain consists in performing a DFT of the record of data captured at the ADC output, in order to calculate the ADC parameters after the corresponding spectral content. In practice, the DFT is usually implemented as an FFT and the number of samples M in the record is a power of two ($log_2 M$ is an integer) because in that case, the algorithm used to perform the FFT is much more simple.

D. Dallet and J. Machado da Silva, (eds.), Dynamic Characterisation of
Analogue-to-Digital Converters, 85–103.

As only real input signals are considered, the frequency spectrum resulting from the DFT at the output of an ADC is symmetrical (even for the amplitude and odd for the phase), that is why only half of the discrete spectrum is considered $\left(k \in [0, \frac{M}{2}]\right)$. Note that both lines $Y[0]$ and $Y[\frac{M}{2}]$ must be included in order to avoid any lack of information.

2.   Choice of input and clock frequencies

When applying a DFT to process the data at the output of an ADC, coherent sampling provides a spectrum that exhibits only frequencies corresponding to the input frequencies and their harmonics . Coherent sampling is defined by

$$M f_i = J f_s \qquad (4.3)$$

where
$$\begin{aligned}
M &= \text{number of samples in the data record,} \\
f_i &= \text{frequency of the input waveform,} \\
J &= \text{integer number of cycles of the input waveform in the data record,} \\
f_s &= \text{sampling frequency.}
\end{aligned}$$

Whenever this relationship is not met, the record is processed using a window weighting function prior to performing the DFT, to minimise spectral leakage effects.

3.   Windowing

Windowing consists in multiplying the data record by a window function $(w[n])$ in the time domain prior to the DFT process. Then the DFT is performed on $w[n]y[n]$,

$$Y[k] = \sum_{n=0}^{M-1} w[n]y[n]\, e^{-i2\pi n \frac{k}{M}} \qquad k = 0 \dots M - 1 \qquad (4.4)$$

As mentioned in section 2, windowing is required when using incoherent sampling. This is due to the fact that the frequency spectrum obtained when a DFT is applied to process M data values sampled at f_s, can only exhibit discrete frequencies with a step of $\frac{f_s}{M}$. If relationship (4.3) is met, the input frequency as well as its harmonics are integer multiples of $\frac{f_s}{M}$. In that case, these frequencies correspond exactly to discrete lines in the spectrum. On the other hand, if (4.3) is not met, the input frequency as well as its harmonics do not correspond to single discrete lines in the spectrum, and thus spectral leakage is observed. Windowing reduces the spectral leakage effects.

3.1   Processing Gain (PG) of a window

Let the input data of a windowed DFT be defined by

$$y[n] = A\, e^{i2\pi n \frac{J}{M}} + \eta[n] \qquad n = 0 \dots M - 1 \qquad (4.5)$$

where $\eta[n]$ is a white noise sequence with variance σ_T^2. The noiseless signal power of the output spectrum is given by the power of the spectral line $Y[J]$,

$$|Y[J]|^2 = A^2 \left[\sum_{n=0}^{M-1} w[n] \right]^2 \qquad (4.6)$$

The output amplitude of the noiseless signal is the input amplitude multiplied by a term which is the sum of the window terms. This term is called the coherent gain of the window.

The noise power of the spectral line $Y[J]$ is given by

$$E\left\{|Y[J]|^2\right\} = \sigma_T^2 \sum_{n=0}^{M-1} w^2[n] \qquad (4.7)$$

where E is the expectation operator. As additive noise is assumed to be white, the value $\sigma_T^2 \sum_{n=0}^{M-1} w^2[n]$ represents the noise floor level (or the noise power spectral density).

The Processing Gain (PG), of the window is defined as the ratio of the output signal to noise floor ratio to the input signal to noise ratio

$$PG = \frac{\dfrac{A^2 \left[\sum_{n=0}^{M-1} w[n] \right]^2}{\sigma_T^2 \sum_{n=0}^{M-1} w^2[n]}}{\dfrac{A^2}{\sigma_T^2}} = \frac{\left[\sum_{n=0}^{M-1} w[n] \right]^2}{\sum_{n=0}^{M-1} w^2[n]} \qquad (4.8)$$

The inverse of PG is called the normalised Equivalent Noise Band Width (ENBW), which is defined by

$$ENBW = \frac{M.\sum_{n=0}^{M-1} w^2[n]}{\left[\sum_{n=0}^{M-1} w[n] \right]^2} \qquad (4.9)$$

The PG of a window as described above neglects the effects resulting from the input frequency value called the picket-fence effect or scalloping loss. For more details, see [71] which also describes the properties of many window functions. Additional window functions with very low sidelobes are presented in [111].

The choice of the window depends on the application but usually, for ADC evaluation, it is preferable to use window functions exhibiting low sidelobes. More detailed explanations on windowing are reported in section 6.9. Note that a non-windowed DFT can be considered as a windowed DFT using a rectangular window ($w[n] = 1$, $n = 0 \ldots M - 1$). The Processing Gain of an M samples long rectangular window is M.

4.   Comment on the accuracy of the input frequency

The aim of this section is to determine the maximum deviation between the ideal input frequency and the real input frequency beyond which windowing is necessary . As explained in section 3, if the coherence relationship (4.3) is not met, spectral leakage is observed. That means that instead of being concentrated in a single line of the frequency spectrum, the power of the input sinewave is spread across all the lines of the spectrum. Thus the inaccuracy of the input frequency of an ADC degrades the measurement of the signal to noise ratio at the output of the ADC because the part of the signal power that is spread across all lines of the spectrum is added to effective noise.

Let the input data of a non-windowed DFT be

$$y[n] = A \, e^{i2\pi n \frac{J \pm \epsilon_j}{M}} + \eta[n] \qquad n = 0 \ldots M - 1 \qquad (4.10)$$

where $\eta[n]$ is a white noise sequence with variance σ_T^2, and ϵ_j is the number of cycles inaccuracy in the input frequency, that is

$$f_i = (J \pm \epsilon_j) \frac{f_s}{M} \qquad (4.11)$$

The signal to noise ratio at the input of the DFT is then

$$SNR_{in} = \frac{A^2}{\sigma_T^2} \qquad (4.12)$$

The noiseless amplitudes of the spectral lines of the spectrum are

$$|Y[k]| = A \, \frac{\sin[\pi(J \pm \epsilon_j - k)]}{\sin\left[\frac{\pi}{M}(J \pm \epsilon_j - k)\right]} \qquad k = 0 \ldots M - 1 \qquad (4.13)$$

From (4.13), if $\epsilon_j = 0$, then the coherence relationship is met and $|Y[k]|^2 = A^2 \times M^2$ for $k = J$ and 0 for $k \neq J$. In that case, all power of the input noiseless signal is concentrated in a single line and the noise power calculated from all the other lines of the normalised spectrum is $\sigma_T^2 \times M^2$. The signal to noise ratio calculated from the spectrum equals the signal to noise ratio at the input of the DFT.

If $\epsilon_j \neq 0$, the coherence relationship is no longer met and the signal power calculated from the line J of the spectrum is

$$|Y[J]|^2 = A^2 \left(\frac{sin(\pi\epsilon_j)}{sin\left(\frac{\pi}{M}\epsilon_j\right)} \right)^2 \tag{4.14}$$

and the noise power is given by

$$\text{noise power} = M^2 \sigma_T^2 + A^2 \left(M^2 - \left(\frac{sin(\pi\epsilon_j)}{sin\left(\frac{\pi}{M}\epsilon_j\right)} \right)^2 \right) \tag{4.15}$$

In that case, due to frequency inaccuracy, the signal to noise ratio calculated from the spectrum is smaller than the signal to noise ratio at the input of the DFT.

The input frequency inaccuracy introduces an error in the calculation of the signal to noise ratio at the output of the DFT. It is possible to calculate the maximum value of ϵ_j and thus the accuracy required for the input frequency to estimate the signal to noise ratio with an error lower than a given value.

Let the error on the signal to noise ratio calculation e be defined by

$$e = \frac{SNR_{in} - SNR_{calculated}}{SNR_{in}} \tag{4.16}$$

where SNR_{in} is defined in (4.12).

The value of ϵ_j that leads to an error e in the calculation of the signal to noise ratio, is determined by solving the following equation

$$\frac{1}{M} \frac{sin(\pi\epsilon_j)}{sin\left(\frac{\pi}{M}\epsilon_j\right)} = \sqrt{\frac{(1-e)(1+SNR_{in})}{1+(1-e)SNR_{in}}} \tag{4.17}$$

Once the value of ϵ_j is determined, given the sampling frequency and the number of samples recorded, the maximum value of the frequency inaccuracy is calculated by

$$\Delta f_{imax} = \pm \frac{\epsilon_j f_s}{M} \tag{4.18}$$

The determination of the maximum input frequency deviation tolerated to measure the SNR of an ideal N-bit ADC with a 0.5 dB error, is given in the following table (in this example, as the ADC is ideal, the signal to noise ratio equals the signal to quantisation error given in (4.26)).

The number of samples used to compute the values in table 4.1 is the power of two immediately greater than the value given in (4.19), that is 2^{N+2}. If the accuracy of the input frequency is not sufficient to allow the calculation of the signal to noise ratio with an error lower than the required value, then windowing is necessary (see section 6.9.3).

Table 4.1. Maximum Δf_i for a 0.5 dB error on the SNR calculation.

N	$\frac{\Delta f_{i\,max}}{f_s}$
4	$\pm 1.5 \times 10^{-4}$
6	$\pm 9.5 \times 10^{-6}$
8	$\pm 6.0 \times 10^{-7}$
10	$\pm 3.7 \times 10^{-8}$
12	$\pm 2.3 \times 10^{-9}$
14	$\pm 1.4 \times 10^{-10}$
16	$\pm 9.1 \times 10^{-12}$
18	$\pm 5.7 \times 10^{-13}$
20	$\pm 3.5 \times 10^{-14}$
22	$\pm 2.1 \times 10^{-15}$
24	$\pm 1.4 \times 10^{-16}$

5.   Record size

5.1   Number of samples required

5.1.1   Deterministic analysis.   The minimum record size required to evaluate an ADC is a record in which there is at least one sample per code bin . For an N-bit ideal ADC in absence of random noise, the minimum record size when applying a full-scale input sine wave is

$$M_{min} = \pi(2^N + 1) \approx \pi\, 2^N \qquad (4.19)$$

When taking into account the ADC differential nonlinearity (DNL), the minimum record size is then

$$M_{min} = \frac{\pi\, 2^N}{1 - |DNL_{max}|} \qquad (4.20)$$

Where $|DNL_{max}|$ is the worst case DNL expressed in LSB.

5.1.2   Probabilistic considerations.   In this section, the total jitter resulting from the ADC, the clock signal and the input signal are taken into account. So sampling can no longer be considered as a deterministic operation and a probabilistic approach must be used. The minimum size of the record is then determined so that it leads to a given probability of acquiring at least one sample per code bin in the record.

For a full scale input sine wave, to have a probability p of having at least one sample in every code bin in a record, the minimum size of the record M_{min}

must be chosen so that

$$\frac{1}{2} \, \text{erfc} \left(\frac{\frac{\pi}{M_{min}} - \frac{1 - |DNL_{max}|}{2^N}}{2\pi f_i \sigma_j} \right) \geq p \tag{4.21}$$

where

erfc = complementary error function,
M_{min} = minimum size of the record,
DNL_{max} = worst case DNL expressed in LSB,
N = the ADC number of bits,
f_i = input frequency,
σ_j = worst case standard deviation of the total jitter.

The complementary error function is defined by

$$\text{erfc}(z) = \frac{2}{\sqrt{\pi}} \int_z^{+\infty} e^{-t^2} dt \tag{4.22}$$

which is related to the error function by

$$\text{erfc}(z) = 1 - \text{erf}(z) \tag{4.23}$$

For more details on the complementary error function, see [11].

5.2   Comment on the required number of samples

The equations given in section 5.1 may result in a very large number of samples in one record for high resolution ADCs. In that case, it may be impossible to acquire the desired number of samples in a single record due to memory depth limitations. This problem can be solved by the acquisition of multiple records, each with a size compatible with the memory depth, associated with averaging in the frequency domain. The first point of each record being random, the number of records needed to reach a given probability of having at least one sample of every code bin is not easily determined theoretically in the case of an input sinewave.

Benkais reports [18] a simulation of the number of records required to reach a probability p of having at least one sample of every code bin as a function of p and of the ratio $\frac{2^N}{M}$ in the case of an input sinewave, where N is the number of bits of the ADC and M is the number of samples in each record. It is shown that, for $N \geq= 10$, to obtain a 95% probability of testing all codes, 280 records should be captured, each comprising $2^N/16$ samples, or 48 records comprising $2^N/4$ samples each.

5.3   Comment on the choice of input and clock frequencies

When coherent sampling is used ($M f_i = J f_s$) with the additional condition that

J is relatively prime to M,

then, there are M distinct samples of the input sine wave on the phase axis. These samples are uniformly distributed between 0 and 2π. So, the use of coherent sampling with M and J mutually prime is the optimum choice of frequency for ADC testing because the number of distinct phases sampled is maximum. If J and M are not mutually prime and if

$$\frac{M}{J} = L, \tag{4.24}$$

then, there are only L distinct phases sampled. In practice, the DFT is usually implemented as an FFT and M is a power of two. Consequently, choosing J to be an odd integer ensures that J and M are mutually prime.

Summary
The use of coherent sampling with a number of cycles of the input waveform in the data record mutually prime to the total number of samples in the data record, avoids spectral leakage and maximises the number of distinct phases sampled. This choice of input and clock frequency is the optimal one.

6.   Calculation of ADC dynamic parameters in the frequency domain

In the first part of this chapter, it is assumed that the input sine wave and clock frequencies are optimum as explained in 5.3

$$f_i = J\frac{f_s}{M} \qquad \text{with } J \text{ and } M \text{ mutually prime.} \tag{4.25}$$

We also consider that a non-windowed DFT is used to process the M data at the output of the ADC. Finally only half of the resulting spectrum $(Y[k]$ with k in $[0; \frac{N}{2}])$ is used for the calculation of the parameters listed in this chapter.

6.1   Quantisation error of an ideal ADC

As shown in [72], for an ideal quantiser the model of an uniformly distributed noise for the quantisation error is valid with a 1 percent accuracy if the quantizer has a number of bits N greater than or equal to 12 . Nevertheless, for most of the quantisers available on the market ($N \geq 6$), the assumption that the quantisation error is a uniformly distributed noise with a variance of $\frac{Q^2}{12}$ does not result in great inaccuracy. That is why, in this section, the quantisation error is considered as a uniformly distributed noise with a white spectrum [147], [136].

For an N-bit ideal ADC, the ratio of the variance of the quantisation error to the rms power of a full scale sine wave is

$$\frac{\sigma^2(\text{quantisation error})}{\left(\frac{V_{fs}}{2\sqrt{2}}\right)^2} = -6.02 \times N - 1.76 \qquad (dBfs) \qquad (4.26)$$

If the DFT performed at the output of this ideal ADC takes into account an M data length non windowed record, the Processing Gain defined in 3.1 is equal to M. If the entire output spectrum is considered, the noise floor to full scale ratio is

$$\text{NFl (dBfs)} = -6.02 \times N - 1.76 - 10\log(M) \qquad (4.27)$$

As mentioned at the beginning of this chapter, we will consider half of the discrete spectrum when performing the calculations. In that case, only half of the rms power of the full scale sine wave is taken into account. So the full scale to noise floor ratio is

$$\begin{aligned}\text{NFl (dBfs)} &= -6.02 \times N - 1.76 - 10\log(M) + 10\log(2) \\ &= -6.02 \times N - 10\log(M) + 1.25\end{aligned} \qquad (4.28)$$

6.2   Noise Floor (NFl) evaluation

As mentioned above, the output spectrum used for the calculation ranges from 0 to $\frac{M}{2}$ inclusive and the ADC input sine wave is located at the spectral line J $\left(f_i = J\frac{f_s}{M}\right)$. In that case, as the ADC's output noise is assumed to be white, the noise floor (or the noise power spectral density) can be calculated by

$$|NFl|^2 = \frac{\displaystyle\sum_{k=1,\,k\neq J,\,k\neq hJ}^{M/2-1} |Y[k]|^2 + \frac{1}{2}\left|Y\left[\frac{M}{2}\right]\right|^2}{\frac{M}{2} - h_{max}} \qquad h = 2\ldots h_{max}$$

$$(4.29)$$

where $Y[hJ]$ is the hth harmonic component and h_{max} is the highest harmonic to remove. Note that $Y[hJ]$ is a notation to represent the hth harmonic component. If $hJ > \frac{M}{2}$, it is the aliased line that must be used in the equation above. To evaluate noise floor , neither the DC line nor the signal line nor the harmonics lines have to be considered. In the equation above, the number of harmonics to remove for the calculation of the noise floor depends on the ADC under test and also on the accuracy required. In practice, it is often sufficient to remove the second through the tenth harmonics. Note also that the power of the $\frac{M}{2}$th spectral line must be divided by 2 because its power represents the total power of the ADC's output signals at frequency $\frac{f_s}{2}$ (for the other spectral lines, as only half of the spectrum is considered, the power of a spectral line represents half of the total power of the ADC output signals at the considered frequency).

6.3   Signal to Noise and Distortion Ratio ($SINAD$)

For a pure sine wave input, $SINAD$ can be calculated from the power spectrum by

$$SINAD_{dB} = 10 \log \frac{|Y[J]|^2 - |NFl|^2}{\displaystyle\sum_{k=1,\,k\neq J}^{M/2-1} |Y[k]|^2 + 2|NFl|^2 + \frac{1}{2}\left|Y\left[\frac{M}{2}\right]\right|^2} \qquad (4.30)$$

As noise is assumed to be white, the term $2|NFl|^2$ takes into account noise in the bins 0 and J.

6.4   Effective Number of Bits (N_{ef})

This is the resolution of an ideal ADC that would have a full-scale signal to quantisation error ratio equal to the $SINAD_{dBfs}$ calculated for the ADC under test

$$N_{ef} = \frac{SINAD_{dBfs} - 1.76}{6.02} \qquad (4.31)$$

where $SINAD_{dBfs}$ is the $SINAD_{dB}$ measured when applying a full-scale sine wave at the input of the ADC.

In practice, as it is difficult to use a full-scale sine wave to measure an ADC's dynamic parameters, a sine wave of amplitude A is used to measure $SINAD_{dB}$, being $SINAD_{dBfs}$ then deduced by

$$SINAD_{dBfs} = SINAD_{dB} - 20 \log (SFSR) \qquad (4.32)$$

where $SFSR$ is the Signal to Full Scale Ratio. This value has to be specified and chosen as close to zero dB as possible to calculate the Effective number of bits of an ADC.

In practice, N_{ef} is often given without specifying the value of $SFSR$, which can lead to a misunderstanding of the performances of the ADC. Moreover, the calculation of N_{ef} does not give any additional information on the ADC comparing to $SINAD$, and actually N_{ef} is not a physical parameter of ADCs. That is why we advise not to use N_{ef} to characterise an ADC.

6.5 Signal to Noise Ratio (SNR)

For a pure sine wave input, the SNR can be calculated from the power spectrum by

$$SNR_{dB} = 10\log \frac{|Y[J]|^2 - |NFl|^2}{\sum\limits_{k=1,\, k\neq J,\, k\neq hJ}^{M/2-1} |Y[k]|^2 + (h_{max}+1)|NFl|^2 + \frac{1}{2}\left|Y\left[\frac{M}{2}\right]\right|^2}$$

$$h = 2 \ldots h_{max}$$

(4.33)

where $Y[hJ]$ is the hth harmonic component and h_{max} is the highest harmonic to remove. Note that $Y[hJ]$ is a notation to represent the hth harmonic component. If $hJ > \frac{M}{2}$, it is the aliased line that must be used in the equation above.

In (4.33), the number of harmonics to be removed for the calculation of the sum in the denominator depends on the ADC under test and also on the accuracy required. This number has to be specified. In practice, it is often sufficient to remove the second through the tenth harmonics. Note that the term $(h_{max}+1)|NFl|^2$ takes into account noise in the 0, J and harmonic bins.

6.6 Total Harmonic Distortion (THD)

This parameter represents the non-linearities of the ADC transfer curve . THD is the ratio of the sum of the squares of all the harmonics including their aliases to the rms power of the fundamental component

$$THD_{dB} = 10\log \frac{\sum\limits_{h=2}^{h_{max}} |Y[hJ]|^2}{|Y[J]|^2}$$

(4.34)

where $Y[hJ]$ is the hth harmonic component. Note that $Y[hJ]$ is a notation to represent the hth harmonic component. If $hJ > \frac{M}{2}$, it is the aliased line that must be used in (4.34).

The number of harmonics used in (4.34) must be specified. If not specified, THD is estimated by using the sum of the squares of the second through the tenth harmonics, inclusive.

6.7 Spurious Free Dynamic Range ($SFDR$)

The definition of a spurious tone given in the terminology is: "spurious tones are persistent sine waves at frequencies f_{sp} included in the range defined as harmonic distortion other than the harmonic frequencies" . In this definition, it is useful to specify what persistent sine waves are: "a persistent sine wave is a line that is at least 10 dBs higher than the noise floor in the averaged power spectrum".

The definition of the averaged power spectrum is: "Let's consider R records of M data at the output of an ADC. For each record, a non windowed FFT is performed which leads to R spectra $Y_1[k] \ldots Y_R[k]$". The averaged spectrum is calculated by

$$|Y_{avm}[k]|^2 = \frac{1}{R} \sum_{i=1}^{R} |Y_i[k]|^2 \qquad (4.35)$$

In practice, the number of records used to calculate the average spectrum and thus to smooth the noise and make the spurious emerge is generally lower than or equal to 10. Taking a number of records greater than 10 will not lead to a great improvement of the noise smoothness.

The $SFDR$ is then defined by

$$SFDR_{dB} = 10 \log \frac{|Y_{avm}[J]|^2}{max_{f_h, f_{sp}} \left\{ |Y_{avm}[f_h]|^2, |Y_{avm}[f_{sp}]|^2 \right\}} \qquad (4.36)$$

where f_h and f_{sp} are, respectively, the frequencies of the set of harmonic and spurious spectral lines present over the full Nyquist band.

6.8   Intermodulation Distortion (IMD)

Intermodulation distortion may occur whenever an ADC is sampling a signal composed of two or more sine waves .

6.8.1   Two tone intermodulation distortion.   Let us assume that two pure sine waves of frequencies f_1 and f_2 are used at the input of an ADC, which are considered to be optimum frequencies, that is

$$
\begin{aligned}
f_1 &= J_1 \frac{f_s}{M} \\
f_2 &= J_2 \frac{f_s}{M}
\end{aligned}
\qquad (4.37)
$$

with J_1 and J_2 relatively prime to M. In that case, the IMD spectral components have coherent frequencies too. Indeed, their frequencies are given by

$$f_{imd} = |jJ_1 + lJ_2| \times \frac{f_s}{M} \qquad (4.38)$$

where j and $l \in \mathbf{Z}$; $|j|$ and $|l| \geq 1$. $|j| + |l|$ is the order of the IMD.

The two tone third order IMD is often specified. In that case, the frequencies of the IMD spectral components are

$$|J_1 \pm 2J_2| \times \frac{f_s}{M} \quad and \quad |J_2 \pm 2J_1| \times \frac{f_s}{M} \tag{4.39}$$

The measurement of the two tone q^{th} order IMD is performed by applying two tones of equal amplitude to the ADC. In practice, the synthesizers used to provide the two tones are not mutually phase locked that is why the power of each tone is set 6 dBs below the power at which the IMD is specified. If no input power is specified, the IMD is the worst case over the full input range of the ADC.

The value of the two tone q^{th} order IMD is given by

$$IMDq_{dB} = 10 \log \frac{min\left\{|Y(J_1)|^2, |Y(J_2)|^2\right\}}{max_{j,l\geq 1, |j|+|l|=q}\left\{|Y[|jJ_1 + lJ_2|]|^2\right\}} \tag{4.40}$$

Where $max_{j,l\geq 1, |j|+|l|=q}\left\{|Y[|jJ_1 + lJ_2|]|^2\right\}$ is the largest power of all the two tone q^{th} order IMD spectral lines over the full Nyquist band.

Note that even if the two input tones have the same amplitude, the amplitudes of the output tones can be slightly different. That is why in (4.40) the numerator is $min\left\{|Y(J_1)|^2, |Y(J_2)|^2\right\}$.

6.9   Changes in the Formulas for Non Coherent Sampling

6.9.1   Effects to take into account.   Windowing is necessary when the coherent sampling relationship (4.3) is not met. In spite of that, a similar relationship between input frequency and clock frequency can also be written for non-coherent sampling

$$f_i = (J \pm \epsilon_j)\frac{f_s}{M} \tag{4.41}$$

where J is an integer and ϵ_j is the number of cycles inaccuracy ($\epsilon_j \leq 0.5$).

When a DFT is performed on an M samples long record, the output spectrum only exhibits discrete lines at integer multiples of $\frac{f_s}{M}$. In the frequency-domain analysis, the amplitude of an input sinewave with a frequency defined as in (4.41) is calculated by the amplitude of line J of the spectrum. As the frequency of the sinewave is not exactly $J.\frac{f_s}{M}$, the determination of its amplitude by using uniquely line J will not lead to a correct result.

Nevertheless, it is possible to calculate the exact value of the amplitude of a sinewave having a frequency defined as in (4.41) by multiplying the amplitude of line J by the following factor

$$\frac{|W[0]|}{|W_c \left(\frac{\epsilon_j f_s}{M} \right)|}$$ (4.42)

where

$$W[0] = \sum_{n=0}^{M-1} w[n]$$

and

$$W_c \left(\frac{\epsilon_j f_s}{M} \right) = \int_{-\infty}^{+\infty} e^{-i2\pi \frac{\epsilon_j f_s}{M} t} w(t) dt$$

$W_c \left(\frac{\epsilon_j f_s}{M} \right)$ can be approximated from $w[n]$, the M samples long window, by adding zeros to $w[n]$ in order to obtain a new sequence $w_L[n]$ that is much longer.
$w_L[n]$ is an L samples long sequence defined as follows

$$\begin{aligned} w_L[n] &= w[n] & \text{for n = 0} \ldots \text{M-1} \\ w_L[n] &= 0 & \text{for n = M} \ldots \text{L-1} \end{aligned}$$ (4.43)

Usually, it is considered sufficient to calculate $|W_c \left(\frac{\epsilon_j f_s}{M} \right)|$ only for values of ϵ_j multiple of 0.1, i.e., for

$$\epsilon_j \in \{0.1; 0.2; 0.3; 0.4; 0.5\}$$

Indeed, for most of the windows, the calculation of the factor defined in (4.42) for values of ϵ_j rounded to one decimal do not differ from its exact value (obtained when ϵ_j is not rounded) by more than 0.3 dB.

For the different equations listed in this section, when using the value of ϵ_j rounded to one decimal, the terminology ϵ_{jr} will be used.

To calculate $|W_c \left(\frac{\epsilon_{jr} f_s}{M} \right)|$, it is sufficient to set the length of the sequence $w_L[n]$ to $10 \times M$. In that case, the value of $|W_c \left(\frac{\epsilon_{jr} f_s}{M} \right)|$ becomes $|W_{10M}[10.\epsilon_{jr}]|$ where $W_{10M}[k]$ is the DFT of $w_L[n]$ performed on the $L = 10.M$ samples of $w_L[n]$.

For the calculation of the dynamic parameters listed in the previous sections, another important effect of windowing is the broadening of the main lobe. Due to that broadening, to calculate noise power, it is necessary to remove, in addition to the signal line and its harmonics, a few bins before and after these lines.

6.9.2   Changes in the formulas.   As the explanations of the different equations are reported above, only the changes to apply to the formulas for non-coherent sampling are given in this section . Due to the main lobe broadening, the first equation that needs to be corrected when windowing is used is the calculation of noise floor

$$|NFl|^2 = \frac{\sum\limits_{k=1,\ k\neq J\pm l,\ k\neq rnd[h(J\pm\epsilon_j]\pm l}^{M/2-1} |Y[k]|^2 + \frac{1}{2}\left|Y\left[\frac{M}{2}\right]\right|^2}{\frac{M}{2} - h_{max}(2l_{max}+1)} \tag{4.44}$$

$$h = 2\ldots h_{max}$$
$$l = 0\ldots l_{max}$$

where $rnd[x]$ is the round to the nearest integer of x, $Y[hJ]$ is the hth harmonic component, $2l_{max}$ is the number of bins to remove around the signal and its harmonics.

The number of bins to be removed depends on the window. For a given window, the value of l_{max} equals the position (in number of bins) of the first zero of the Fourier Transform of the window (l_{max} equals half of the width of the main lobe). The value of l_{max} is reported for some classical windows in section 6.9.3.

In addition to the main lobe broadening and to signal amplitude correction, the ENBW of the applied window must be used to correct the equation provided in section 4.6.5 for the SNR calculation

$$SNR_{dB} = 10\log$$

$$\frac{|Y[J]|^2 - |NFl|^2}{\sum\limits_{\substack{k=1 \\ k\neq J\pm l \\ k\neq rnd[h(J\pm\epsilon_j]\pm l}}^{M/2-1} |Y[k]|^2 + (h_{max}(2l_{max}+1)+1)|NFl|^2 + \frac{1}{2}\left|Y\left[\frac{M}{2}\right]\right|^2}$$

$$+ 10\log(ENBW) + 10\log\left(\frac{|W[0]|^2}{\left|W_c\left(\frac{\epsilon_{jr}f_s}{M}\right)\right|^2}\right)$$

$$\tag{4.45}$$

with $h = 2\ldots h_{max}$ and $l = 0\ldots l_{max}$, and $rnd[x]$ being the round to the nearest integer of x. The calculation of $W_c\left(\frac{\epsilon_{jr}f_s}{M}\right)$ is explained in section 6.9.1.

ENBW is defined in section 4.3.1 as

$$ENBW = \frac{M.\sum\limits_{n=0}^{M-1} w^2[n]}{\left[\sum\limits_{n=0}^{M-1} w[n]\right]^2} \tag{4.46}$$

The same corrections are necessary for the calculation of SINAD from a windowed DFT. To perform a rigorous correction, it is necessary to split the sum used in (4.30) in order to separate the noise from the harmonics. After a windowed DFT, SINAD is calculated by

$$
\begin{aligned}
SINAD_{dB} =& 10\log\frac{|Y[J]|^2 - |NFl|^2}{A+B} + 10\log(ENBW) \\
&+ 10\log\left(\frac{|W[0]|^2}{|W_c\left(\frac{\epsilon_{jr}f_s}{M}\right)|^2}\right)
\end{aligned}
\tag{4.47}
$$

where

$$A = \sum_{k=1,\, k\neq J\pm l,\, k\neq rnd[h(J\pm\epsilon_j]\pm l}^{M/2-1} |Y[k]|^2 + (2l_{max}+2)|NFl|^2 + \frac{1}{2}\left|Y\left[\frac{M}{2}\right]\right|^2 \tag{4.48}$$

with $l = 0\ldots l_{max}$ and

$$B = ENBW \sum_{h=2}^{hmax} |Y[rnd[h(J\pm\epsilon_j)]]|^2 . \frac{|W[0]|^2}{|W_c\left(\frac{frac_r[h(J\pm\epsilon_j)]f_s}{M}\right)|^2} \tag{4.49}$$

where $rnd[x]$ is the round to the nearest integer of x, and $frac_r[x]$ is the fractional part of x rounded to the first decimal.

Note that to calculate SINAD, the amplitude of the signal as well as the amplitudes of the harmonics have to be corrected. Similarly to what has been done for the signal amplitude, the correction factors used for the harmonic lines are only defined for values rounded to one decimal.

In the calculation of THD, the signal and harmonics amplitudes must be modified due to non-coherent sampling

$$THD_{dB} = 10\log \frac{\sum\limits_{h=2}^{h_{max}} |Y[hJ]|^2 \dfrac{|W[0]|^2}{|W_c\left(\frac{frac_r[h(J\pm\epsilon_j)]f_s}{M}\right)|^2}}{|Y[J]|^2 \dfrac{|W[0]|^2}{|W_c\left(\frac{\epsilon_{jr}f_s}{M}\right)|^2}}$$

$$= 10\log \left(\frac{\sum\limits_{h=2}^{h_{max}} \dfrac{|Y[hJ]|^2}{|W_c\left(\frac{frac_r[h(J\pm\epsilon_j)]f_s}{M}\right)|^2}}{|Y[J]|^2} \cdot |W_c\left(\frac{\epsilon_{jr}f_s}{M}\right)|^2 \right)$$

(4.50)

where $frac_r[x]$ is the fractional part of x rounded to the first decimal.

For the calculation of SFDR, only the signal amplitude can be corrected because the exact frequency of spurious lines are generally unknown. The equation to be used to calculate SFDR in the case of non-coherent sampling is

$$SFDR_{dB} = 10\log \frac{|Y_{avm}[J]|^2}{max_{f_h, f_{sp}}\left\{ |Y_{avm}[f_h]|^2, |Y_{avm}[f_{sp}]|^2 \right\}}$$

$$+ 10\log \left(\frac{|W[0]|^2}{|W_c\left(\frac{\epsilon_{jr}f_s}{M}\right)|^2} \right)$$

(4.51)

For the calculation of the two tones IMD in the case of non-coherent sampling, two tones having the following frequencies are considered

$$f_1 = (J_1 \pm \epsilon_{j1})\frac{f_s}{M} \qquad \text{and} \qquad f_2 = (J_2 \pm \epsilon_{j2})\frac{f_s}{M}$$

(4.52)

In that case, the amplitude of the tone under consideration and the amplitude of the intermodulation line taken into account for the calculation of the qth

order IMD must be modified

$$IMDq_{dB} = 10 \log \left(\frac{min\left\{|Y(J_1)|^2, |Y(J_2)|^2\right\} \frac{|W[0]|^2}{|W_c\left(\frac{\epsilon_{j(1,2)r}f_s}{M}\right)|^2}}{max_{j,l\geq 1, |j|+|l|=q}\left\{|Y[|jJ_1 + lJ_2|]|^2\right\}} \right.$$

$$\left. \times \frac{1}{\frac{|W[0]|^2}{|W_c\left(\frac{frac_r[j(J_1\pm\epsilon_{j1})+l(J_2\pm\epsilon_{j2})]f_s}{M}\right)|^2}} \right) \qquad (4.53)$$

$$= 10 \log \frac{min\left\{|Y(J_1)|^2, |Y(J_2)|^2\right\}}{max_{j,l\geq 1, |j|+|l|=q}\left\{|Y[|jJ_1 + lJ_2|]|^2\right\}}$$

$$+ 10 \log \frac{|W_c\left(\frac{frac_r[j(J_1\pm\epsilon_{j1})+l(J_2\pm\epsilon_{j2})]f_s}{M}\right)|^2}{|W_c\left(\frac{\epsilon_{j(1,2)r}f_s}{M}\right)|^2}$$

where $W_c\left(\frac{\epsilon_{j(1,2)r}f_s}{M}\right)$ equals $W_c\left(\frac{\epsilon_{j1r}f_s}{M}\right)$ if $|Y(J_1)|^2 < |Y(J_2)|^2$ and $W_c\left(\frac{\epsilon_{j2r}f_s}{M}\right)$ otherwise. $frac_r[x]$ is the fractional part of x rounded to the first decimal.

6.9.3 Selection of the window.

The selection of the window depends on the resolution of the ADC to be characterised. The higher the resolution is, the lower the side-lobes of the window have to be. Nevertheless, lowering the side-lobes results in increasing the main lobe width.

For some measurements configuration (two tones of very close frequencies for example), a large increase of the main lobe width may not be tolerable and the selection of the window is a compromise between the needed side-lobe reduction and a tolerable increase in main lobe.

For SNR measurements, low side-lobes are necessary in order to lower the leakage, and thus to increase SNR calculation accuracy. A simulation of the error on the SNR calculation for some classical windows allows the window to be chosen as a function of the resolution of ideal ADCs. For that simulation, the worst case of the number of cycles inaccuracy was considered ($\epsilon_j = 0.5$)

$$f_i = (J + 0.5)\frac{f_s}{M} \qquad (4.54)$$

As only ideal ADCs are considered, the SNR at the input of the DFT is

$$SNR_{dB} = 6.02N + 1.76 \qquad (4.55)$$

Table 4.2. Value of l_{max} for some classical windows.

Window	l_{max}
Hanning (W1)	2
Hamming (W2)	2
Blackman (W3)	3
Exact Blackman (W4)	3
Windows of fig.15 in [111] (W5)	4
7 term Blackman-Harris (W6)	7

Table 4.3. Error in SNR calculation for some classical windows as a function of the number of bits.

N	W1	W2	W3	W4	W5	W6
4	-0.8	-1	-0.2	-0.2	$> -10^{-1}$	$> -10^{-1}$
6	-6	unusable	-2.4	-2.5	$> -10^{-1}$	$> -10^{-1}$
8	unusable	"	unusable	unusable	$> -10^{-1}$	$> -10^{-1}$
8	"	"	"	"	$> -10^{-1}$	$> -10^{-1}$
10	"	"	"	"	$> -10^{-1}$	$> -10^{-1}$
12	"	"	"	"	-0.6	$> -10^{-1}$
14	"	"	"	"	-5.3	$> -10^{-1}$
16	"	"	"	"	unusable	$> -10^{-1}$
18	"	"	"	"	"	$> -10^{-1}$
20	"	"	"	"	"	$> -10^{-1}$
22	"	"	"	"	"	$> -10^{-1}$
24	"	"	"	"	"	-0.13

where N is the ADC number of bits.

The SNR calculation is performed using (4.45) (simplified as there are no harmonics) and the result is compared to the SNR at the input of the DFT.

When using (4.45), the value of l_{max} has to be specified. The table bellow lists the values of l_{max} used in (4.45) for some classical windows

Table 4.3 lists the error on the SNR calculation when using (4.45) comparing to the SNR of the ideal ADC. The names of the windows Wi in this table are identified in Table 4.2. The error is defined as $10 \log \left(\frac{SNR_{calculated}}{SNR_{in}} \right)$. In this table, a window is said "unusable" when the absolute value of the error is greater than 6 dB. Note that the number of samples used to perform the DFT is not mentioned because the error on the SNR calculation is independent of this parameter.

Chapter 5

CODE HISTOGRAM TEST

Giovanni Chiorboli

Dip. di Ingegneria dell'Informazione, University of Parma
Parco Area delle Scienze 181/A, 43100 Parma, Italy

giovanni.chiorboli@unipr.it

Carlo Morandi

Dip. di Ingegneria dell'Informazione, University of Parma
Parco Area delle Scienze 181/A, 43100 Parma, Italy

carlo.morandi@unipr.it

1.   Introduction

The histogram test is based on the assumption that the sampled input signal V is a random variable whose distribution function is accurately known. Let $p\{V\}$ represent the probability density function (p.d.f.) of V, and

$$P\{V\} = \int_{-\infty}^{V} p[\xi]d\xi$$

the corresponding distribution function. Then, the position of code transition level $T[k]$, separating code bins $k-1$ and k, may be determined by considering that the probability of obtaining a code smaller than k can be experimentally estimated by the relative frequency Φ_r of the event "output code $< k$",

$$\mathcal{P}rob\{\text{output code} < k\} = P\{T[k]\} = \int_{-\infty}^{T[k]} p[\xi]d\xi \approx \Phi_r \tag{5.1}$$

and therefore $T[k]$ may be determined as

$$T[k] = P^{-1}\{\Phi_r\}, \tag{5.2}$$

assuming that the inverse function of P exists.

D. Dallet and J. Machado da Silva, (eds.), Dynamic Characterisation of
Analogue-to-Digital Converters, 105–156.
© 2005 Springer. Printed in the Netherlands.

For static testing, an input signal with uniform p.d.f. over the input range of the converter, such as a triangular waveform, is sometimes preferred in view of the easier data processing. For dynamic testing, a high purity sinewave is preferred for two reasons: the first is that low distortion and low-noise sinewave generators are readily found in the market and it is relatively easy, by filtering, to improve distortion and noise to the desired level; the second is that it is easy to quantitatively assess the purity of the sinewave by means, for instance, of a spectrum analyzer. The input waveform may therefore be represented by

$$v(t) = A \, \cos(\omega t + \phi) + C \, , \tag{5.3}$$

where $A > 0$ is the amplitude, $\omega = 2\pi f_i$ is the angular frequency, $f_i = T_i^{-1}$ is the frequency, ϕ is the initial phase and C the offset. A and C are normally chosen so as to span all the code transition levels of the ADC.

If the sampling phase is a random variable uniformly distributed in $[0, 2\pi)$, the probability $P\{V\}$ of collecting a sample with a value between $C - A$ and V, where $C - A \leq V \leq C + A$, is represented by the fraction of the period 2π in which $v(t) \leq V$. In order to estimate this fraction, it may be assumed without loss of generality that $\phi = 0$. The inequality

$$A \, \cos(\omega t) + C \leq V \tag{5.4}$$

in $[0, 2\pi)$ is satisfied, see Figure 5.1, for

$$\arccos\left(\frac{V - C}{A}\right) \leq \omega t \leq 2\pi - \arccos\left(\frac{V - C}{A}\right) \tag{5.5}$$

i.e. in an interval of half width $\psi(V) = \left(\pi - \arccos(\frac{V-C}{A})\right) = \arccos(\frac{C-V}{A})$ centered on π. Thus $P\{V\}$, the distribution function of V, is

$$P\{V\} = \frac{2 \, \psi(V)}{2\pi} = \frac{1}{\pi} \arccos\left(\frac{C - V}{A}\right). \tag{5.6}$$

Let us assume that A and C are chosen so as to span all the codes of the ADC. Then, if $V = T[k]$, where $T[k]$ is the kth code transition level, separating code bins $k - 1$ and k, the probability of obtaining an output code smaller than k is

$$\mathcal{P}rob\{\text{output code} < k\} = P\{T[k]\} = \frac{1}{\pi} \arccos\left(\frac{C - T[k]}{A}\right). \tag{5.7}$$

Taking the inverse

$$T[k] = C - A \, \cos(\pi \, \mathcal{P}rob\{\text{output code} < k\}) = C - A \cos \psi_k, \tag{5.8}$$

where $\psi_k = \psi(T[k])$. Let us consider an experiment where a large number of samples is acquired at random, uniformly distributed phases. Let $h[i]$ be

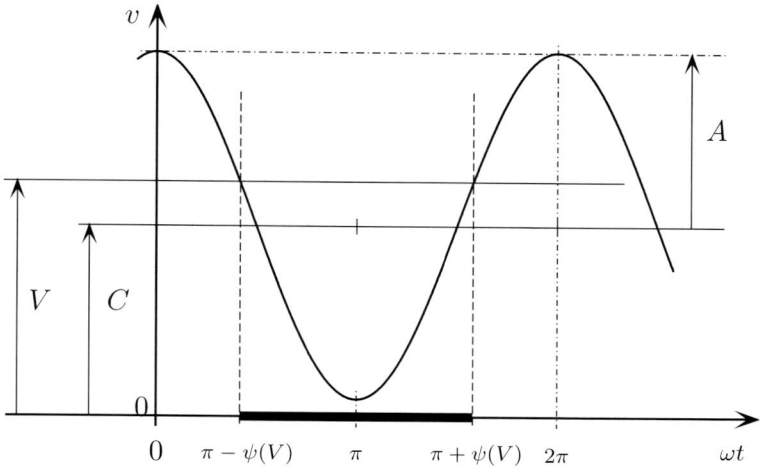

Figure 5.1. The region where inequality (5.4) is satisfied is marked by the thick line.

the total number of samples which yielded output code i as the result of the conversion (the so called *"code histogram"*), and $ch[k] = \sum_{i=0}^{k} h[i]$ represent the cumulative histogram. For an N bit binary ADC, the total number of samples collected will be $S = \sum_{i=0}^{2^N-1} h[i]$. Then, replacing probability $\mathcal{P}rob\{\text{output code} < k\}$ by the relative frequency observed during the test, $ch[k-1]/S$, we estimate

$$\hat{T}[k] = C - A \cos \left(\frac{\pi \, ch[k-1]}{S} \right). \tag{5.9}$$

It may be shown that (5.9) is an asymptotically unbiased estimator of $T[k]$, whose bias is negligible for all values of M employed in practice [29]. The above formula is based on the assumption that the relative frequency is a good estimate of $\mathcal{P}rob\{\text{output code} < k\}$, that the distribution function of V is represented by (5.6) and that the sampling phases are random and uniformly distributed in $[0, 2\pi)$. However, the relative frequency $ch[k]/S$ may be an inaccurate estimate of $\mathcal{P}rob\{\text{output code} < k\}$ due to the finite size of the sample, synchronous or quasi-synchronous deterministic sampling may be preferred to random sampling and, finally, additive noise and distortion modify the distribution function of the input signal. All these points will be reviewed in the following sections.

Once the code transition levels are determined, it is easy to obtain all the specification parameters which may be derived from the knowledge of the conversion characteristic, such as gain error, offset, integral or differential non linearity.

Provided that the amplitude and offset of the input sinewave are precisely known, the conversion characteristic obtained from the histogram test using a low-frequency sinewave usually matches well the conversion characteristic measured at DC, for instance by the servo-loop arrangement.

At the highest frequencies, however, a degradation of the conversion characteristic is usually observed, revealing the rise of dynamic non-idealities.

Clearly, if this happens, revealing that the result of the conversion depends not only on the input signal, but also on the internal state of the ADC, we cannot expect that a conversion characteristic determined from a sinewave of given frequency, amplitude and offset may describe the ADC operation in response to any other different input stimulus. Nevertheless, the results obtained may be useful for comparing the behaviour of different ADCs subject to the same stimulus.

2.   The sampling strategy and its contribution to count variance and measurement uncertainty

2.1   Random vs coherent sampling

The approach based on sampling the input sinewave at random phases [48], uniformly distributed in $[0, 2\pi)$, is discussed in subsection 2.2. It turns out that a very large number of samples is required in order to achieve an acceptable measurement uncertainty, and therefore the approach may be recommended only when synchronization is not possible.

The required number of samples could be substantially reduced if, on the contrary, it were possible to perfectly synchronize the input sinewave with the sampling clock, so as to obtain a set of samples equally spaced by phase increments so small, that the number of samples falling within the less populated code bin is sufficient to determine its width with the required uncertainty. The case of perfect synchronization is discussed in subsection 2.3.

A more realistic analysis, considering synchronous sinewave and clock, but accounting for frequency ratio inaccuracy [28] is carried out in section 2.4.

2.2   The case of random sampling

The basic assumption is that the sampling phases are uniformly distributed in $[0, 2\pi)$.

Figure 5.2, quite similar to Figure 5.1, apart from a $-\pi$ shift in the horizontal axis, shows a possible arrangement of the sampling phases. If the phase ψ of the sample, reduced to the range $[-\pi, \pi)$, falls in the interval $-\psi_k, \psi_k$, the sample contributes to the counts in $ch[k-1]$. Clearly, the probability for a sample to fall in $[-\psi_k, \psi_k]$ is

$$p = \frac{\psi_k}{\pi},$$ (5.10)

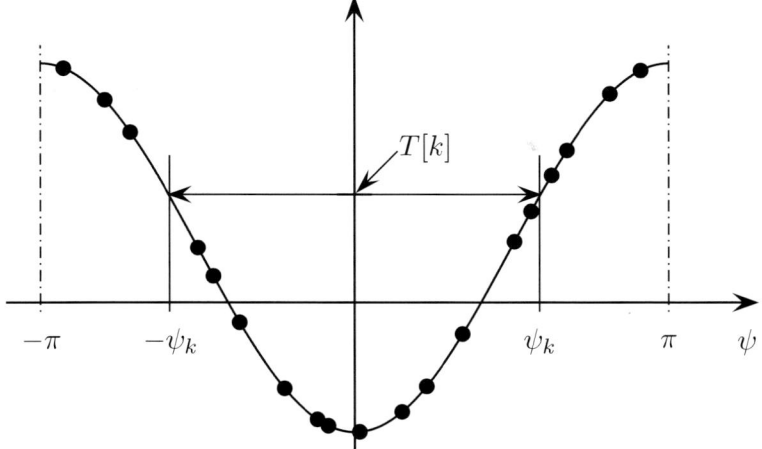

Figure 5.2. Determination of the phase interval $[-\psi_k, \psi_k]$ from the transition level $T[k]$. The dots represent a set of sampling phases in the case of random sampling.

while $1 - p$ is the probability of falling outside. If S samples are extracted, the probability that m of them fall in $[-\psi_k, \psi_k]$, so that $ch[k-1] = m$, is represented by the binomial distribution

$$P\{m\} = \binom{S}{m} p^m (1-p)^{S-m}, \tag{5.11}$$

Thus, $ch[k-1]$ is a random variable with expected value and variance given by

$$E\{ch[k-1]\} = pS \tag{5.12}$$
$$\sigma^2_{c,intr} = Sp(1-p), \tag{5.13}$$

where the subscript *"c,intr"* stands to remember that it is a contribution to count variance intrinsically arising from the counting process. From (5.9),(5.12) and (5.13) it is then possible to infer the expected value and variance of the estimates of the transition levels.

The expected value of $\hat{T}[k]$ is $E\{\hat{T}[k]\} = C - A\cos(\psi_k)$.

As for the contribution to the variance of $\hat{T}$ arising from the intrinsic count uncertainty, $\sigma^2_{T_{c,intr}}$, by differentiating (5.9)

$$\frac{d\,\hat{T}}{d\,ch} = \frac{\pi A}{S} \sin\left(\frac{\pi\,ch[k-1]}{S}\right), \tag{5.14}$$

and hence

$$\sigma^2_{T_{c,intr}} = \left(\frac{\pi A}{S}\right)^2 \sin^2\left(\frac{\pi ch[k-1]}{S}\right) \quad \sigma^2_{c,intr} \leq \frac{(\pi A)^2}{S} p(1-p) \leq \frac{(\pi A)^2}{4S} \tag{5.15}$$

since the maximum of $p(1-p)$ is $1/4$.

If the number of samples S is large, the binomial distribution may be approximated by a gaussian distribution with the same mean and standard deviation. It is then easy to determine an interval about the mean value $\mu = E\{\hat{T}[k]\}$ which contains, with confidence $1 - \alpha$, the transition level $T[k]$. It is sufficient to determine a number $Z_{\alpha/2}$ such that

$$\int_{\mu-Z_{0,\alpha/2}\sigma}^{\mu+Z_{0,\alpha/2}\sigma} \frac{1}{\sigma\sqrt{2\pi}} \exp\left(-\frac{(\xi-\mu)^2}{2\sigma^2}\right) d\xi = 1 - \alpha, \qquad (5.16)$$

where $\sigma = \sigma_{T_{c,intr}}$ and $Z_{0,\alpha/2}$ can be easily determined using tabulated values of the error function erf [11].

Let Q be the nominal code bin width (1 LSB). If an extended uncertainty (tolerance) BQ is required with confidence $1 - \alpha$, then

$$Z_{0,\alpha/2}\, \sigma_{T_{c,intr}} \leq BQ \qquad (5.17)$$

hence, from the last inequality in (5.15)

$$S \geq \frac{Z_{0,\alpha/2}^2 \pi^2 A^2}{4B^2 Q^2}\,. \qquad (5.18)$$

Note that a very large number of samples is required. When the input signal spans the input range, (5.18) becomes

$$S \geq \frac{Z_{0,\alpha/2}^2 \pi^2 2^{2(N-1)}}{4B^2}\,: \qquad (5.19)$$

for an 8 bit converter, an uncertainty of 0.1 LSB and 99 % confidence level, S shall be greater than $2.7\ 10^7$.

Finally note that (5.19) ensures only that the estimate $\hat{T}[k]$ of a single code transition level does not differ from the expected value $\mu = E\{\hat{T}[k]\}$ by more than BQ with confidence $(1-\alpha)$. If it is required that none of the code transition levels of an N-bit converter differs from the corresponding expected value by more than BQ with confidence $(1-\alpha)$, $Z_{0,\alpha/2}$ must be replaced by $Z_{N,\alpha/2}$, as discussed later in section 5.6.

2.3   The case of perfectly coherent sampling

In coherent sampling , the S samples used to build the frequency histogram are obtained from a single record acquired at the constant sampling rate $f_s = \frac{1}{T_s}$. The input sinewave is sampled in correspondence of S distinct values of phase, which appear equally spaced by $\Delta\phi = \frac{2\pi}{S}$ when they are rearranged in the $[-\pi, \pi)$ interval. Note that the samples may be collected either in real time, by choosing a sampling frequency f_s which is precisely a multiple, by the factor S, of the signal frequency f_i, or in equivalent time. In the last case, which is analyzed in the Appendix, Figure 5.14, the signal is carefully synchronized to the sampling clock, so that in each record of S samples there is an integer number J of cycles of the input sinewave, and J is mutually prime to S.

The resulting situation is depicted in Figure 5.3: the width of the phase

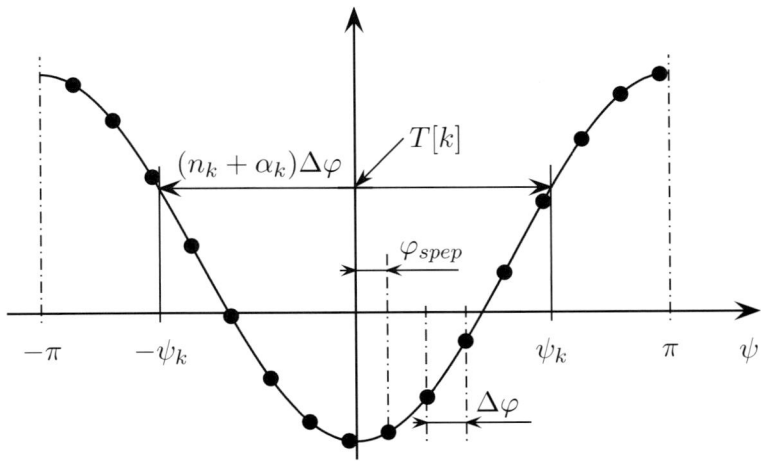

Figure 5.3. Determination of the phase interval $[-\psi_k, \psi_k]$ from the transition level $T[k]$. The dots represent the sampling phases in the case of coherent sampling

interval $[-\psi_k, \psi_k]$ is measured by counting the number of equally spaced sampling phases contained therein. The problem is thus reduced to evaluating the uncertainty associated with the measurement of the length of a segment using a regulus with marks regularly spaced by $\Delta\phi$.

When the cumulative histogram is formed, the number of counts $ch[k-1]$ falling in the cell corresponding to $T[k]$ is the number of samples with phase angles between $-\psi_k$ and ψ_k. The relative position of the sampling comb with respect to the interval $[-\psi_k, \psi_k]$ can be defined by ϕ_{spep}, the smallest positive equivalent-phase angle among the sampling points. It is assumed that ϕ_{spep} is a random variable uniformly distributed between 0 and $\Delta\phi$; this is the same as requiring that the initial phase in a record is a random variable with uniform distribution in $[0, 2\pi)$.

If the length of the phase interval $[-\psi_k, \psi_k]$ is written in the form $2\psi_k = (n_k + \alpha_k)\Delta\phi$, where n_k is an integer and $0 < \alpha_k < 1$, the probability of counting $(n_k + 1)$ marks is α_k, the probability of counting n_k is $(1 - \alpha_k)$. If the random variable *"extra count"* x is defined, which assumes the value 1 if $n_k + 1$ is counted, 0 if n_k is counted, it is apparent that x obeys a point binomial distribution [105], so that the expected value of the count is $n_k + \alpha_k$, and the count variance is $\sigma^2_{c,intr} = \alpha_k(1 - \alpha_k)$.

Now, α_k depends on the code transition level: since we are interested in obtaining estimates of uncertainty applicable to any transition level, we must consider the maximum value of count variance, which occurs for $\alpha_k = 0.5$ and equals 0.25.

The corresponding variance affecting the estimated transition level $\hat{T}[k]$, is obtained as in the previous section (see (5.14) and following).

$$\sigma^2_{T_{c,intr}} = \left(\frac{dT}{dch}\right)^2 \sigma^2_{c,intr} \leq \frac{1}{4}\left(\frac{\pi A}{S}\right)^2. \tag{5.20}$$

Comparing (5.20) and (5.15) it is apparent that the r.h.s. of (5.20) is smaller by a factor S, quite relevant. However, achieving adequate coherence between the sampling clock and the sinusoidal signal is quite difficult in practice.

2.4   Quasi-coherent sampling

When the test signal frequency becomes comparable to the sampling frequency, it is only possible to resort to equivalent time sampling . Records containing M samples are acquired at the constant sampling rate $f_s = \frac{1}{T_s}$. The input sinewave, with frequency f_i, is carefully synchronized to the sampling clock, so that in each record of M samples there is an integer number J of cycles of the input sinewave, and J is mutually prime to M. Normally, several (R) records are acquired to build the histogram, so that $S = RM$. We shall consider here the case where the histogram is formed using one single record, $R = 1$.

In practice, the coherence condition $r \triangleq f_i/f_s = J/M$, where J and M are mutually prime integers, is never perfectly met, because of the finite resolution of the synthesizers and because of the lowest frequency components of phase noise. It is therefore convenient to represent the frequency ratio as:

$$\frac{f_i}{f_s} = \frac{J}{M} + \Delta r, \tag{5.21}$$

where Δr is the fractional frequency deviation.

As before, the task is evaluating the length of a segment using a ruler; this time, however, its marks are unequally spaced according to the sampling pattern resulting from (5.21). The initial phase in a record is assumed as a random variable uniformly distributed in $[-\pi, \pi)$.

Figure 5.4 shows the sampling pattern in a quasi-coherent case defined by $J = 3$, $M = 16$ and $\Delta r = 1/(2JM)$.

The problem was thoroughly analyzed in [28]. Leaving as usual $2\psi_k = (n_k + \alpha_k) 2\pi/M$, with n_k integer and $0 < \alpha_k < 1$, the probabilities $p_i \triangleq$

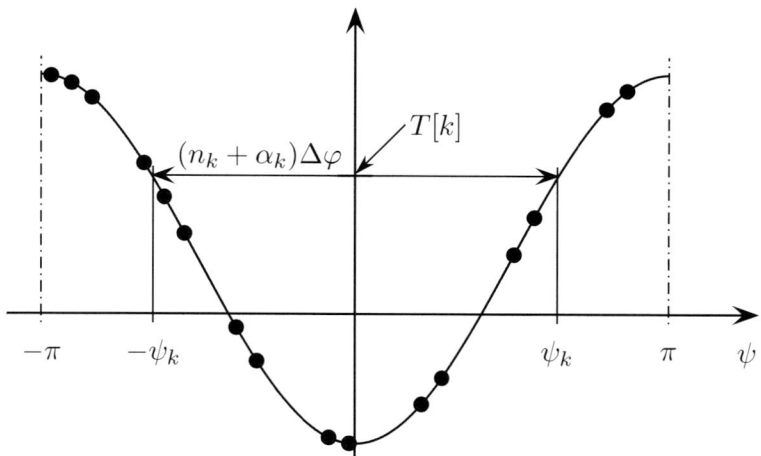

Figure 5.4. Determination of the phase interval $[-\psi_k, \psi_k]$ from the transition level $T[k]$. The dots represent a set of possible sampling phases in the case of quasi-coherent sampling

$Prob\{ch[k-1] = n_k + i\}$ can be derived as a function of ψ_k and of f_i/f_s, as shown in the appendix. By using these probabilities, the expected value and the variance of $ch[\cdot]$ are also evaluated in the Appendix. It turns out that

$$E\{ch[k-1]\} = n_k + \alpha_k, \tag{5.22}$$

while the count variance is given in Table 5.1.

In the table m_L (m_R) is the positive (respectively, negative) solution of the equation

$$mJ \bmod M = n_k, \quad -M < m < M \tag{5.23}$$

where mod represents the modulus operator. Similarly, n_L and n_R are the positive, respectively negative, solutions of

$$nJ \bmod M = n_k + 1 \quad -M < n < M. \tag{5.24}$$

Closed form expressions for m_L, m_R, n_L and n_R are derived in the appendix. It turns out that, if

$$\frac{\Delta r}{r} \le \frac{1}{2JM} \tag{5.25}$$

the variances reported in Table 5.1 are bounded by $1/4$, as it was found in the case of perfectly coherent sampling. If condition (5.25) is not respected, the maximum variance can be determined by numerical simulation, as reported in [13]. So, if condition (5.25) is respected, even if a frequency-ratio error Δr is present, the variance affecting the estimated transition level $\hat{T}[k]$ does not increase beyond the value reported in (5.20) for perfectly coherent sampling,

Table 5.1. Variance of the number of counts. The initial phase of the sinewave is assumed uniformly distributed in $[-\pi, \pi)$. (a) $0 \le \Delta r < 1/(2M^2)$; (b) $-1/(2M^2) < \Delta r < 0$.

(a)	$0 \le \alpha_k < Mm_L\Delta r$	$Mm_L\Delta r \le \alpha_k < 1+Mn_R\Delta r$	$1+Mn_R\Delta r \le \alpha_k < 1$
$\sigma^2_{c,intr}$	$\alpha_k\left(\frac{m_R+m_L}{M}-\alpha_k\right)-2m_Rm_L\Delta r$	$\alpha_k(1-\alpha_k)$	$\alpha_k(1-\alpha_k)+2\frac{n_L}{M}[\alpha_k-1-Mn_R\Delta r]$

(b)	$0 \le \alpha_k < Mm_R\Delta r$	$Mm_R\Delta r \le \alpha_k < 1+Mn_L\Delta r$	$1+Mn_L\Delta r \le \alpha_k < 1$
$\sigma^2_{c,intr}$	$2m_Rm_L\Delta r-\alpha_k\left(\frac{m_R+m_L}{M}+\alpha_k\right)$	$\alpha_k(1-\alpha_k)$	$\alpha_k(1-\alpha_k)-2\frac{n_R}{M}[\alpha_k-1-Mn_L\Delta r]$

i.e.

$$\sigma_{T_{c,intr}}^2 \leq \frac{\pi^2 A^2}{4M^2} \tag{5.26}$$

For the validity of this result, no particular relationship between J and M is required, except that J and M must be mutually prime. So, the result can be applied even well beyond the Nyquist limit.

Finally, if it is desired that each record contains at least one sample in each code bin, it is worth to mention that for an assigned record length M, equal to a power of two, tables are provided in [25] which report values of J particularly tolerant of errors ΔJ either in the positive direction ($\Delta J > 0$) or in the negative direction ($\Delta J < 0$), together with a search strategy that may help in selecting the test frequency.

3. Additional contributions to count uncertainty: additive noise and jitter

3.1 Count uncertainty associated with additive noise

When additive noise is superimposed to the input sinewave, a sample point whose phase falls within the interval $[-\psi_k, \psi_k]$ corresponding to transition level $T[k]$, and is near one of the extremes of the interval (this means that the sampled voltage, in the absence of noise, would be slightly lower than $T[k]$) has a non negligible probability of bringing no contribution to $ch[k-1]$. And vice-versa. In the following discussion, along the line of [23], it is assumed that noise has a Gaussian distribution with zero mean and standard deviation σ_n.

3.1.1 Contribution to transition level uncertainty.
Let us consider a sample, acquired at such phase that, in the absence of noise, it would be classified in the cell $ch[k-1]$ of the cumulative histogram. In the presence of noise, will it - or not - be classified in that cell?

If the voltage that would be sampled in the absence of noise is $T[k]-x$, with $x > 0$, then, in the presence of noise, that particular sample will be classified in cell $ch[k-1]$ as long as the instantaneous noise amplitude is smaller than x. That is, the probability $p(x)$ that the acquired sample is classified in the right cell, $ch[k-1]$, is

$$p(x) = \frac{1}{\sigma_n \sqrt{2\pi}} \int_{-\infty}^{x} e^{-\frac{v^2}{2\sigma_n^2}} dv \overset{\triangle}{=} 1 - \frac{1}{2}\text{erfc}\left(\frac{x}{\sigma_n\sqrt{2}}\right). \tag{5.27}$$

Let us associate, to each individual sampling point i, a random variable b_i which represents the contribution of that sampling point to the counts in $ch[k-1]$ and is defined by

$b_i = 1$ if the sample is classified in cell $ch[k-1]$;
$b_i = 0$ if the sample is classified in cell $ch[k]$ or in higher cells.

Then, the mean value of the count contribution b_i is $p(x)$ and the variance is $\sigma_b^2(x) = p(x)(1 - p(x))$.

Conversely, if the voltage that would be sampled in the absence of noise is greater than $T[k]$ by the amount $x > 0$, then that sampling point will contribute to $ch[k - 1]$ if the instantaneous noise value is below $-x$. In this case, the mean value of the count contribution b_i is $(1 - p(x))$ and the variance is again $\sigma_b^2(x) = p(x)(1 - p(x))$.

The variance, as it may readily be verified from (5.27), is function only of the ratio $\xi = x/\sigma_n$, and may be plotted in normalized form as a function of ξ, as shown in Figure 5.5.

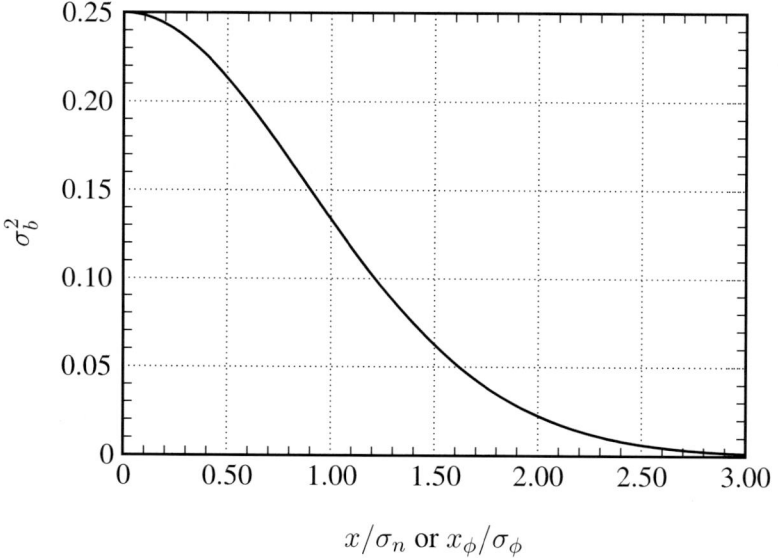

$$x/\sigma_n \text{ or } x_\phi/\sigma_\phi$$

Figure 5.5. The variance in count σ_b induced by the additive noise (by the phase jitter) affecting a sample point at a distance x from $T[k]$ (respectively, at a distance x_ϕ from ψ_k). The abscissa is normalized to the standard deviation of the random variable: x is normalized to σ_n in the case of additive noise, x_ϕ is normalized to σ_ϕ in the case of jitter.

Now, the contributions to count variance of all the sampling points should be added. However, as shown in Figure 5.5, only the sampling points within a few σ_n from $T[k]$ bring an appreciable contribution, i.e. only the sampling points which correspond to sampling phases belonging to the four regions immediately inside or immediately outside the two extremes of the interval $[-\psi_k, \psi_k]$.

If λ (approximately constant) is the number of sampling points per unit voltage in these four regions, then the total contribution to count uncertainty is

$$\sigma_{c,n}^2 = 4\lambda \int_0^\infty \sigma_b^2(x)dx. \tag{5.28}$$

In order to convert this to a variance in voltage, considering that an increment in counts has to be divided by 2λ in order to obtain the corresponding increment in voltage (the points at both extremes of the interval bring their contribution), $\sigma_{c,n}^2$ has to be divided by $4\lambda^2$.

λ may be calculated considering that $M/2\pi$ sampling points per unit phase are available, in the case of perfectly coherent sampling, and that roughly the same number is available in conditions of quasi-coherency. Considering that to an assigned phase interval $\Delta\Phi$ around the instantaneous phase $\Phi = \omega t + \phi$, a voltage interval of width $A\sin(\Phi)\Delta\Phi$ may be associated, we find:

$$\lambda = \frac{M}{2\pi} \frac{1}{A\sin(\Phi)} \geq \frac{M}{2\pi A}. \tag{5.29}$$

Considering also that

$$\int_0^\infty \sigma_b^2(x)dx = \sigma_n \int_0^\infty \sigma_b^2(\xi)d\xi = 1.13\frac{\sigma_n}{4}, \tag{5.30}$$

the last equality resulting from numerical evaluation of the integral. The variance of the estimate of the code transition level becomes therefore:

$$\sigma_{T,n}^2 \leq \frac{1.13}{2} \frac{\sigma_n A\pi}{M}. \tag{5.31}$$

3.1.2 Contribution to code bin width uncertainty. If the width of a code bin is sensibly larger than the standard deviation of additive noise, it is reasonable to assume that the errors affecting the estimates of the positions of the two code transition levels which delimit the bin are independent random variables. As a consequence, the variance of the width estimate will be twice the variance affecting code transition levels.

For converters with medium-high resolution, however, it frequently happens that the standard deviation of additive noise is larger than the code bin width, so that the assumption of independence no longer holds.

Following [23], it will be shown that in this case, twice the variance affecting code transition levels is a rather pessimistic assumption.

Let us consider, for simplicity, a code bin of width equal to the nominal width Q, centered at voltage V_{bin}. With reference to Figure 5.6, consider the case where, in the absence of noise, the voltage V_i would be sampled. Let $v = V - V_i$ represent the additive noise associated with the sample corresponding to the nominal sampling voltage V_i, and let $P\{v\}$ represent the p.d.f. of noise, with zero mean and standard deviation larger than the code bin width Q. The probability that a sampling point corresponding to the nominal sampling voltage V_i falls in the considered bin may be approximated by $p \approx QP\{V_{bin} - V_i\}$. As in the previous section, the contribution of this specific sampling point to

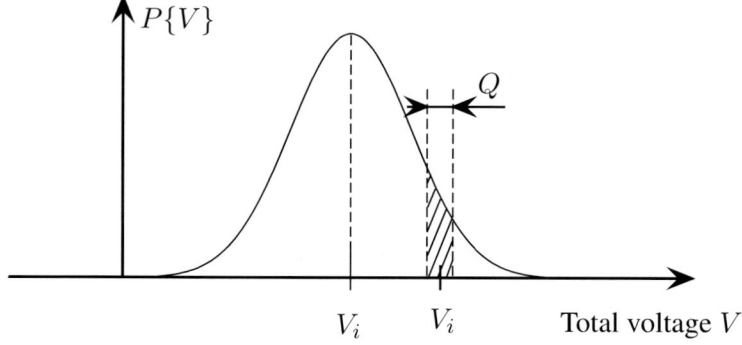

Figure 5.6.   P.d.f. of the noise associated with the nominal sampling voltage V_i.

the count variance will be $p\,(1-p) \leq p$. The total count variance may be evaluated by summing the contributions of all the sampling points.

$$\sigma^2_{c,n} \leq \sum_i QP\{V_{bin} - V_i\} \tag{5.32}$$

By multiplying and dividing the r.h.s. by Δv, the average distance between the sampling voltages near the considered code bin,

$$\sigma^2_{c,n} \leq \frac{Q}{\Delta v} \sum_i P\{V_{bin} - V_i\}\Delta v \approx \frac{Q}{\Delta v} \tag{5.33}$$

since the sum is approximately equal to the area of the p.d.f., i.e. unity.

Considering that $\Delta v = \frac{1}{2\lambda}$, where λ is defined by (5.29) and that the factor 2 arises from the presence of two contributions, one for increasing and the other for decreasing input voltages, then

$$\sigma^2_{c,n} \leq 2\lambda Q \tag{5.34}$$

Considering that in any case, for code bins much larger than the standard deviation of noise, the variance is twice the one estimated for code transition levels, so that for large code bins $\sigma^2_{c,n} \leq 2\lambda\,1.13\,\sigma_n$, it turns out that in general

$$\sigma^2_{c,n} \leq 2\lambda \min\{Q, 1.13\sigma_n\}. \tag{5.35}$$

In terms of code bin width, this yields

$$\sigma^2_{w,n} \leq \frac{\sigma^2_{c,n}}{4\lambda^2} \leq \frac{\pi A}{M}\min\{Q, 1.13\sigma_n\}. \tag{5.36}$$

3.2   Uncertainty associated with sampling jitter

3.2.1   Sources of jitter.   The phase-noise of a signal source is usually described by the single-side bandwidth (SSB) phase-noise spectral power density $\mathcal{L}(f_o)$ (measured in dBc/Hz) as a function of the offset f_o from the carrier, see Figure 5.7 . For each signal generator, the variance σ_θ^2 of the short-term

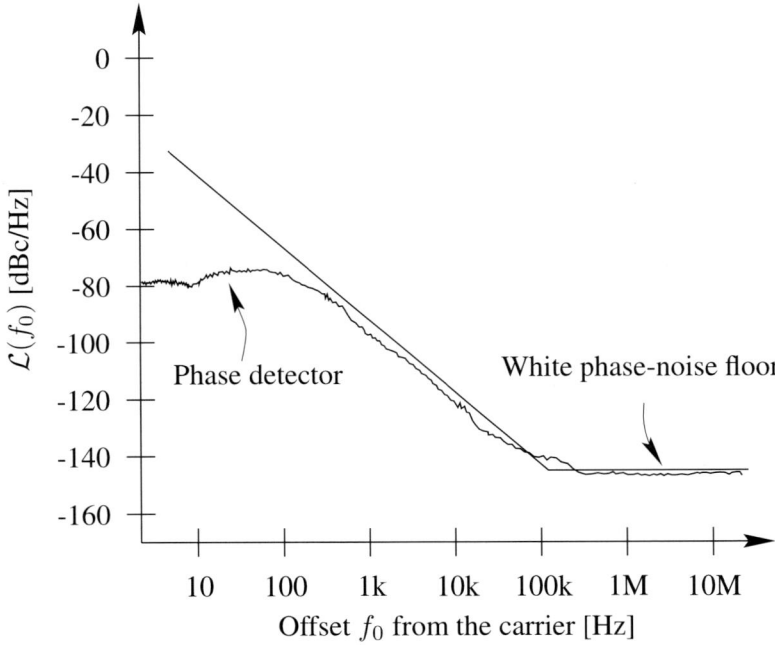

Figure 5.7.   Typical SSB phase noise spectral power density of a sinewave generator.

phase fluctuations should be evaluated, since it results in a nonuniform spacing (in real-time) of the samples acquired in a single record. A quite rough estimate of σ_θ^2, in the case where a selective bandpass filter with 3dB bandwidth BW centered on the synthesized frequency is placed before the converter input, is provided by[1]

$$\sigma_\theta^2 \simeq 2 \int_{f_L}^{f_H} \mathcal{L}(f_o)df_o, \tag{5.37}$$

where $f_L = f_s/M$ and $f_H = BW/2$. When the bandpass filter is not present or is not properly centered, part of the phase noise is transformed into amplitude noise [31], and the evaluation of σ_θ^2 should be carried out accordingly.

Short-term phase fluctuations in the two generators, together with the intrinsic aperture uncertainty of the ADC under test, contribute to the worst case

---

[1] see equations (22.9) and (22.10) in [38].

phase jitter, whose variance is σ_ϕ^2. Let $\sigma_{\theta_{sig}}^2$ be the phase noise variance of the signal generator, operating at frequency f_i; $\sigma_{\theta_{ck}}^2$ the phase noise variance of the sampling clock generator, operating at frequency f_s; σ_{apu}^2 the variance of the aperture delay (aperture uncertainty) of the ADC under test, which may be determined as described in [10, 33]. By referring the total jitter to the input signal frequency one obtains

$$\sigma_\phi^2 = \left(\sigma_{\theta_{sig}}^2 + \frac{\sigma_{\theta_{ck}}^2 f_i^2}{(f_s)^2} + \sigma_{apu}^2 (2\pi f_i)^2 \right) \tag{5.38}$$

Note that other parts of the test setup may contribute to jitter, and in particular the waveform shapers used to obtain the digital clock from the synthesized generator and the signal amplifiers; the above equation can therefore be used only if such contributions are negligible.

3.2.2   Effects of jitter on count variance.   In the presence of jitter a sample that would normally be classified in the cell $ch[k-1]$ of the cumulative histogram which corresponds to $T[k]$, has a finite probability of being not classified in that cell.

Let ϕ_k be the phase angle, within the interval $[-\pi, \pi)$, which would correspond to sample k in the absence of jitter, and let $\delta\phi_k$ represent the deviation from the jitter-free sampling phase ϕ_k.

We shall assume that the probability density function (p.d.f.) describing the phase jitter $\delta\phi_k$ is the same gaussian distribution $P\{\delta\phi\}$ for any k, and that σ_ϕ is the associated standard deviation.

Figure 5.8 shows $P\{\delta\phi\}$ for a few sampling points near the right margin of the interval $[-\psi_k, \psi_k]$, while Figure 5.9 shows the position of the sampling points in real time.

Figure 5.8 demonstrates that in the case $J \geq 2$ the standard deviation of phase jitter may become comparable to the equivalent sampling interval $\Delta\Phi = \frac{2\pi}{M}$, while Figure 5.9 shows that still the natural order of the samples in real time is respected. It may well happen that the actual positions of two adjacent samples in equivalent phase are reversed, i.e.it is possible, e.g., that sample 15 corresponds to a higher voltage than sample 26, while there is no risk of such an exchange between sample 25 and sample 26, since jitter is, by definition, a fluctuation around the mean value of the time interval between two successive samples. In such conditions, it may be assumed that there is negligible correlation between each sample and its nearest neighbours in equivalent phase: for example, in the case shown, two adjacent samples are displaced by about $\frac{MT_s}{3}$ in real time, and at such distance little correlation is expected between them. Note however that when sampling near the Nyquist limit ($J \simeq M/2$),

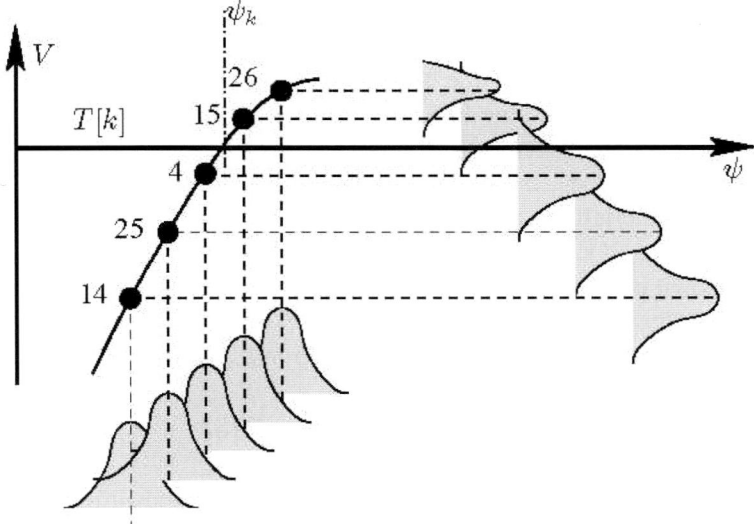

Figure 5.8. The figure shows how phase noise is converted to voltage noise, and that the labelling of adjacent points in equivalent phase corresponds to widely spaced samples in real time (see also Figure 5.9).

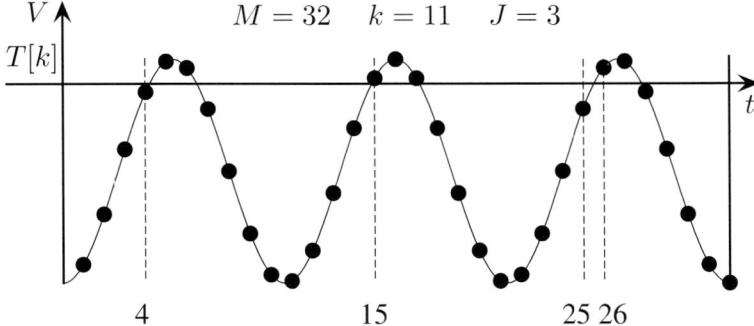

Figure 5.9. The figure shows, for the same case considered in Figure 5.8, the position in real time of the sample points.

the distance in real time between two samples adjacent in equivalent phase is so short, that the above assumption may become questionable.

Thus, considering that only those few sample points within three standard deviations from ψ_k contribute to the count variance (see Figure 5.5), points which are most likely uncorrelated for the reasons explained above, the same theory developed in section 3.1.1 can be applied to evaluate the contribution to count variance.

Let $p(x)$ represent now the probability that a sample, whose position in the absence of jitter is near one edge of the phase interval $[-\psi_k, \psi_k]$, at a distance x_ϕ from that edge is recorded in the cumulative histogram cell $ch[k-1]$ corresponding to $T[k]$.

Following exactly the same line of thought as in section 3.1.1, with reference, however, to the extremes of the phase interval $[-\psi_k, \psi_k]$ rather than to the transition level $T[k]$, the count variance contributed by a sampling point immediately at the left (or at the right) of ψ_k, at a distance x_ϕ from ψ_k is, as before, $\sigma_b^2 = p(x_\phi)(1 - p(x_\phi))$, and the dependence of σ_{bn}^2 on $\xi = x_\phi/\sigma_\phi$ is the same shown in Figure 5.5 after suitable relabeling of the abscissa.

Since $\frac{M}{2\pi}$ is the average density of sampling points per unit phase, the count variance, considering the four contributing regions, becomes

$$\sigma_{c,j}^2 = 4\frac{M}{2\pi} \int_0^\infty \sigma_b^2(x_\phi)dx_\phi = 1.13M\frac{\sigma_\phi}{2\pi}. \qquad (5.39)$$

In order to obtain the corresponding variance in the code transition voltage, the above value is divided by $4\lambda^2$, λ being defined by (5.29).

$$\sigma_{T,j}^2 = \frac{\sigma_{c,j}^2}{4\lambda^2} = \frac{1}{4}A^2\sin^2\psi_k \left(\frac{2\pi}{M}\right)^2 \frac{1.13M\sigma_\phi}{2\pi} \le 1.13A^2\frac{\pi\sigma_\phi}{2M}. \qquad (5.40)$$

4.   Factors affecting the p.d.f. of the input signal

4.1   Modification of the input signal p.d.f. due to additive noise and resulting requirements on overdrive

In this section it is shown that, by applying an input sinewave of sufficient amplitude to trespass, by a convenient amount, the extreme code transition levels $T[1]$ and $T[2^N - 1]$, the systematic error, arising from the difference between the probability density function of the sinewave alone (assumed in the theory) and the p.d.f. of sinewave plus additive noise, can be made negligible .

4.1.1   A useful approximation.   Let

$$v(t) = A \cos(\omega t + \phi) + C \qquad (5.41)$$

be the input signal, which is assumed to cross all the code transition levels of the ADC under test. A normalized input signal x can be defined as

$$x(t) = \frac{v(t) - C}{A}; \quad -1 \le x \le 1. \qquad (5.42)$$

If the r.m.s. value of the additive noise is σ_n in input units, then in normalized units it becomes $\sigma_x = \frac{\sigma_n}{A}$.

Following [23], it is convenient to preliminary proof the following.

If $f[x]$ is an arbitrary function with continuous second derivative, and $p[x]$ is a probability density function (p.d.f.) with zero mean and standard deviation σ_x, then the convolution of $f[x]$ with $p[x]$, $f[x] * p[x]$, can be approximated by

$$f[x] * p[x] \approx f[x] + \frac{\sigma_x^2}{2} f''[x]. \tag{5.43}$$

Proof

    by Taylor expansion of $f[t]$ around $t = x$ up to the second order terms

$$f[t] \approx f[x] + f'[x](t - x) + \frac{1}{2} f''[x](t - x)^2. \tag{5.44}$$

Therefore

$$
\begin{aligned}
f[x] * p[x] &= \int_{-\infty}^{+\infty} f[t] p[x - t] dt \\
&\approx f[x] \int p[x - t] dt + f'[x] \int (t - x) p[x - t] dt \\
&\quad + \frac{1}{2} f''[x] \int (t - x)^2 p[x - t] dt .
\end{aligned} \tag{5.45}
$$

The first integral is one, since $p[x]$ is a probability density function; the second is zero because the mean value of the p.d.f. is zero, the third is the variance σ_x^2. Thus (5.43) is obtained.

4.1.2   Overdrive required for the evaluation of code bin widths.   The probability density function $g[x]$ for signal plus noise is the convolution of the p.d.f. $f[x]$ of the signal with the p.d.f. $p[x]$ of noise [114] . Therefore, according to (5.43),

$$g[x] = f[x] * p[x] \approx f[x] + \frac{\sigma_x^2}{2} f''[x]. \tag{5.46}$$

When measuring the width W of a code bin in position x, the measured code bin width W_m is proportional to the number of samples falling in the code bin, which in turn is proportional to the p.d.f. $g[x]$ at the position x of the bin. If W is the code bin width which would be measured in the absence of noise (p.d.f. $f[x]$), then

$$\frac{W_m}{W} \approx \frac{g[x]}{f[x]} \tag{5.47}$$

and therefore the relative systematic error E_{Wpdf} is

$$E_{Wpdf} \equiv \frac{W_m - W}{W} \approx \frac{g[x]}{f[x]} - 1 \approx \frac{\sigma_x^2}{2} \frac{f''[x]}{f[x]}. \tag{5.48}$$

For the normalized sinewave, the p.d.f. may be obtained by derivation of the distribution function $F[x] = (\arccos(-x))/\pi$ obtained from (5.6) by converting to normalized units.

$$f[x] = \frac{dF}{dx} = \frac{1}{\pi\sqrt{1-x^2}}. \tag{5.49}$$

Hence,

$$f''[x] = \frac{1+2x^2}{\pi(1-x^2)^{5/2}} \tag{5.50}$$

and

$$E_{Wpdf} \approx \frac{\sigma_x^2}{2} \frac{1+2x^2}{(1-x^2)^2} = \frac{\sigma_x^2}{2} \frac{1+2x^2}{(1+x)^2(1-x)^2}. \tag{5.51}$$

This is an even function of x, with vertical asymptotes in $x = \pm 1$. Since $-1 \leq x \leq 1$, the worst case can be studied by considering (5.51) near $x = 1$, replacing x with 1 everywhere but in $(1-x)$:

$$E_{Wpdf} \approx \frac{3\sigma_x^2}{8(1-x)^2}. \tag{5.52}$$

If E_{Wpdf} is sufficiently small, its contribution to the overall uncertainty, when it is combined quadratically with the other uncertainty sources, can be neglected.

To this aim, for a given maximum admitted systematic error E_{Wpdf}^0, it is sufficient to choose x so that

$$(1-x) \geq \sigma_x \sqrt{\frac{3}{8E_{Wpdf}^0}} \tag{5.53}$$

which corresponds, returning to non normalized quantities, to

$$(A+C-v) \geq \sigma_n \sqrt{\frac{3}{8E_{Wpdf}^0}}. \tag{5.54}$$

Considering that $A + C$ is the maximum value of the input sinewave, and that v is an arbitrary value defining the position of the code bin, and therefore a value contained within the input range of the ADC under test, the above inequality must hold also for the code bin near the most positive code transition level $T[2^N - 1]$. Considering the definition of positive overdrive voltage $V_{OD} = A + C - T[2^N - 1]$, E_{Wpdf}^0 will not be exceeded for any code bin if

$$V_{OD} \geq \sigma_n \sqrt{\frac{3}{8E_{Wpdf}^0}}. \tag{5.55}$$

which demonstrates that, in order to reduce to the desired level the systematic error, it is sufficient to adequately increase the overdrive. Note that a negative overdrive voltage of comparable amplitude is necessary to ensure that the widths of the code bins near $T[1]$ are measured with negligible error. However, (5.55) is based on approximation (5.46), i.e. on neglecting terms of order higher than two in the Taylor expansion. This approximation is questionable when x approaches ± 1, and indeed an error larger than the assigned E^0_{Wpdf} may be observed in such cases, even if condition (5.55) is respected. The convolution $f[x] * p[x]$ was therefore calculated numerically for a uniform and a triangular p.d.f. of noise, for values of σ_x between 10^{-2} to 10^{-4}, and for gaussian noise down to $\sigma_x = 10^{-6}$.

Figure 5.10 compares, for different values of σ_x, the true behaviour of $g[x]$ near $x = -1$ and the approximate expression reported in (5.46). When the distance from -1 is larger than $3\sigma_x$, the ratio between the true and the approximate expression of $g(x)$ is always smaller than 1.44, and tends to 1 as the number of standard deviations σ_x increases, so that (5.55) tends to be valid. In any case, the error decreases for increasing overdrive.

In fact, it is convenient to refer directly to figure 5.11, which shows the exact calculation of $E_{Wpdf} \equiv g/f - 1$. As the figure shows, at a distance $3\sigma_x$ from -1, the relative error is $E_{Wpdf} = 5\%$, for all the values of σ_x, and drops to 1% at about $6\sigma_x$.

4.1.3 Overdrive required for the evaluation of code transition levels.

If $F[x]$ is the distribution function of the input sinewave, i.e. the probability that the input signal is $\leq x$, and $p[x]$ is the p.d.f. of noise (all in normalized units), then the distribution function of signal plus noise, $G[x]$, is the convolution of $F[x]$ and $p[x]$:

$$G[x] = \int_{-\infty}^{\infty} p[\xi] F[x - \xi] d\xi = F[x] * p[x] \approx F[x] + \frac{\sigma_x^2}{2} F''[x] \qquad (5.56)$$

where the last approximation is derived from (5.43), and σ_x is the standard deviation of the noise. For the normalized sinewave

$$F[x] = \frac{\arccos(-x)}{\pi} \text{ and } F''[x] = \frac{x}{\pi(1 - x^2)^{3/2}}. \qquad (5.57)$$

If x is the correct position of a code transition level, the position x_m which will be calculated using (5.8) is

$$\begin{aligned}
x_m &= -\cos(\pi G[x]) \\
&\approx -\cos\left(\pi F[x] + \frac{\pi}{2}\sigma_x^2 F''[x]\right) \\
&\approx -\cos(\pi F[x]) + \sin(\pi F[x]) \frac{\pi \sigma_x^2 F''[x]}{2},
\end{aligned} \qquad (5.58)$$

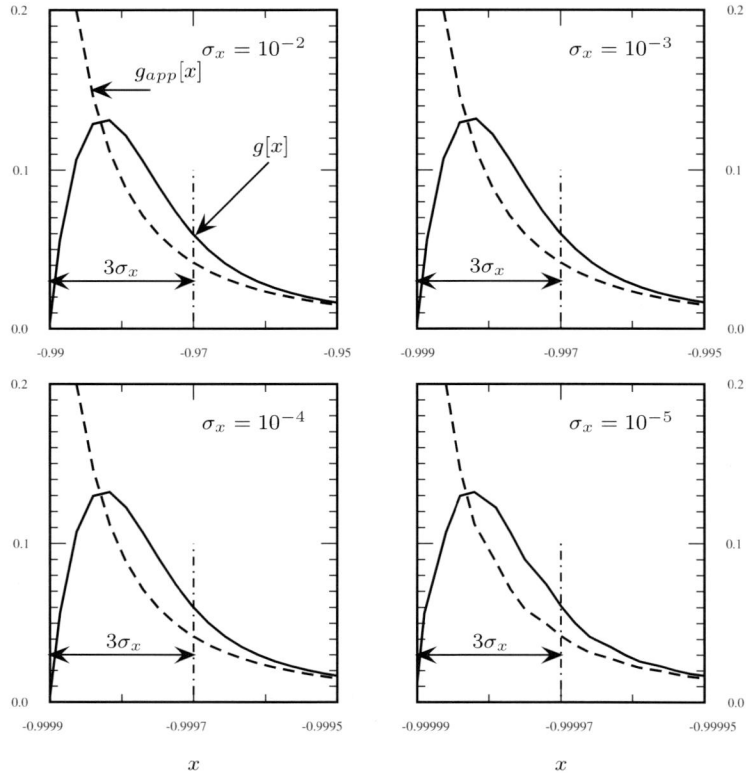

Figure 5.10. The true $g[x]$ compared with the approximation $g_{app}[x] = f[x] + \frac{\sigma_x^2}{2} f''[x]$, see (5.46). The case of gaussian noise is considered here, which in all cases gave the largest error.

where the last was obtained by Taylor expansion. By substituting (5.57), recalling that $-\cos(\pi F[x]) = x$ and $\sin(\pi F[x]) = \sqrt{1 - x^2}$ it turns out

$$x_m - x \approx \frac{\sigma_x^2 x}{2(1 - x)(1 + x)} \tag{5.59}$$

which becomes

$$x_m - x \approx \frac{\sigma_x^2}{4(1 - x)} \tag{5.60}$$

near $x = 1$. As in the previous section, it is desired that the error, arising from the modification of the distribution function due to noise, is smaller than an assigned number E^0_{INLpdf} of code bin widths. This implies

$$1 - x \geq \frac{\sigma_x^2 2^{N-1}}{4 E^0_{INLpdf}} \tag{5.61}$$

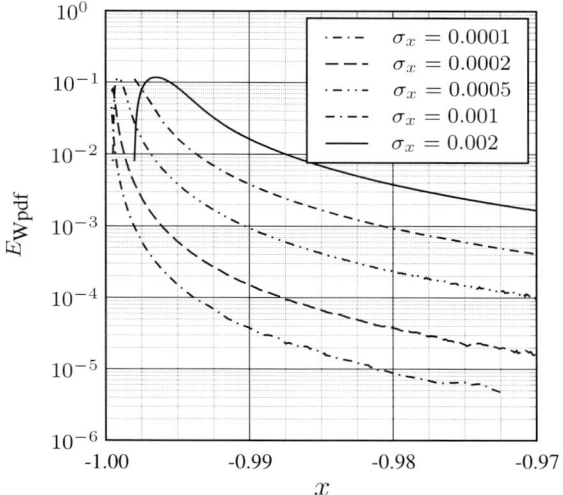

Figure 5.11. The relative systematic error as determined by numerical integration.

Returning to non normalized variables, and remembering that $V_{OD} = A + C - T[2^N - 1]$, with the same considerations reported in the previous section the following constraint is obtained:

$$V_{OD} \geq \frac{\sigma_n^2 2^{N-1}}{4AE_{INLpdf}^0} \approx \frac{\sigma_n^2 2^N}{4E_{INLpdf}^0 V_{rir}}, \tag{5.62}$$

where $V_{rir} \equiv T[2^N - 1] - T[1]$ is the reduced input range of the converter.

As in the previous section, even if condition (5.62) is respected, the actual error may become larger than E_{INLpdf}^0 when x approaches ± 1. According to [23], if the exact convolution is calculated numerically in the cases of gaussian or uniform noise, for values of σ_x between 10^{-2} and 10^{-4}, it turns out that, for an overdrive of $2\sigma_n$ the maximum error E_{INLpdf} does not exceed 1.28 times the error predicted by 5.60.

Thus, in order to obtain an error smaller than an assigned E_{INLpdf}, it is sufficient to require that

$$V_{OD} \geq \max\left(2\sigma_n, \frac{1.28\sigma_n^2}{4E_{INLpdf}Q}\right). \tag{5.63}$$

4.2 Effects of input signal distortion

4.2.1 Introduction. Let

$$v(t) = A \cos(\omega t + \phi) + C \tag{5.64}$$

represent the ideal sinewave, and let

$$v_D(t) = A \, \cos(\omega t + \phi) + C + \sum_{i=2}^{h} A_i \, \cos(i\omega t + \phi_i) \tag{5.65}$$

be the distorted input signal, where $h-1$ harmonic terms are considered ($A_i >$ 0). A value $h = 10$ is normally acceptable .

If the distorting term $d(t)$ obeys

$$|d(t)| = \left| \sum_{i=2}^{h} A_i \, \cos(i\omega t + \phi_i) \right| \le \varepsilon, \tag{5.66}$$

the probability of collecting a sample with a value smaller than or equal to V, $P_D\{V\}$, is bound by

$$P\{V - \varepsilon\} \le P_D\{V\} \le P\{V + \varepsilon\}, \tag{5.67}$$

where $P\{V\}$ is the distribution function for the undistorted signal, defined by (5.6).

In fact, as shown in Figure 5.12, assuming a uniform distribution of the sampling instants, $P_D\{V\}$ is represented by the ratio between the time interval where $v_D(t) < V$ (horizontal solid line) and the sinewave period T_i. Considering that the distorted waveform $v_D(t)$ is always contained in a ribbon 2ε wide around the undistorted waveform $v(t)$, the width of this interval can vary at most between $P\{V - \varepsilon\}$, the interval where $v(t) < V - \varepsilon$ marked by the dash-dotted line, and $P\{V + \varepsilon\}$, the interval where $v(t) < V + \varepsilon$ marked by the dashed line. Both $P\{V\}$ and $P_D\{V\}$ are strictly monotonic, increasing functions of V, since $v_D(t)$ is a single-valued function of t. Thus, $P\{V\}$ and $P_D\{V\}$ can be inverted.

4.2.2   Effects of distortion on INL estimates.   When, in the presence of harmonic distortion, the transition levels are estimated from (5.9), a systematic error has to be accounted for, contributing to the uncertainty of INL and DNL measurements . If $T[k]$ is the correct value of the kth transition level, then the probability that a code smaller than k is collected is:

$$p_k = P_D\{T[k]\} \approx \frac{ch[k-1]}{S} \tag{5.68}$$

and, due to (5.67),

$$P\{T[k] - \varepsilon\} < p_k < P\{T[k] + \varepsilon\}. \tag{5.69}$$

The correct value of $T[k]$ can in principle be derived from p_k using the inverse function P_D^{-1}: $T[k] = P_D^{-1}\{p_k\}$. However, since the inverse function of P

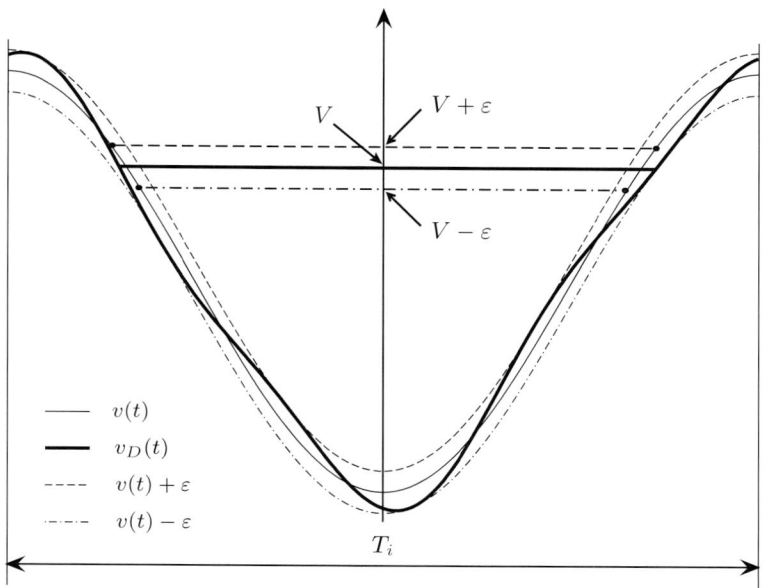

Figure 5.12. $P_D\{V\}$ is represented, in units of T_i, by the thick horizontal segment, which is bound between $P\{V - \varepsilon\}$, dash-dotted segment, and $P\{V + \varepsilon\}$, dashed segment.

is used in (5.9), what is actually estimated is $\hat{T}[k] = P^{-1}\{p_k\}$. Now, since the inverse function of a strictly monotonic, increasing function is a strictly monotonic, increasing function, from the inequality

$$P\{T[k] - \varepsilon\} \le p_k \le P\{T[k] + \varepsilon\} \tag{5.70}$$

it follows

$$P^{-1}\left(P\{T[k] - \varepsilon\}\right) \le P^{-1}\left(p_k\right) \le P^{-1}\left(P\{T[k] + \varepsilon\}\right) \tag{5.71}$$

i.e.

$$T[k] - \varepsilon \le P^{-1}\{p_k\} \le T[k] + \varepsilon. \tag{5.72}$$

The difference between the estimated and the true T_k,

$$h(p_k) = P^{-1}\{p_k\} - P_D^{-1}\{p_k\} \tag{5.73}$$

is therefore bounded between $-\varepsilon$ and ε. Since the relative phases of the harmonics are normally unknown, a conservative approach is to replace ε by the worst-case value, $\sum A_i$.

Therefore, if $E_{INLdist}$ represents, in code bin widths, the maximum admitted systematic error affecting the considered transition level, it must be

$$\varepsilon \le \sum_{i=2}^{h} A_i \le E_{INLdist}Q. \tag{5.74}$$

For instance, in order to evaluate a transition level with an error below 0.1LSB for an 8 bit converter, the ratio $\sum A_i/A$ must be smaller than -62dBc, or smaller than -86dBc for a 12 bit ADC.

4.2.3   Effects of distortion on DNL estimates.   For what concerns DNL , the error induced by distortion is represented by

$$|h(p_{k+1}) - h(p_k)| = |h'(\hat{p})| \, (p_{k+1} - p_k) \qquad (5.75)$$

where the r.h.s. is obtained by Rolle's theorem, $\hat{p}$ being a suitable value between p_k and p_{k+1}.The problem is now reduced to developing upper bounds for the two terms at the r.h.s. of (5.75).

For what concerns $h'(\hat{p})$, let us first remark that $h(p) = P^{-1}\{p\} - P_D^{-1}\{p\}$ represents the reconstruction error $\hat{T}[\cdot] - T[\cdot]$. So

$$|h'(\hat{p})| = \left| \frac{d}{dp} \left(P^{-1}\{p\} - P_D^{-1}\{p\}\right) \right|_{p=\hat{p}} \equiv \left| \frac{d\hat{T}[\cdot]}{dp} - \frac{dT[\cdot]}{dp} \right|_{p=\hat{p}} . \qquad (5.76)$$

The two derivatives have to be evaluated for the same value of probability, $\hat{p}$. Now, it is much easier to evaluate the derivatives of the inverse functions, $\frac{dp}{d\hat{T}[\cdot]}$ and $\frac{dp}{d\hat{T}[\cdot]}$, making reference to Figure 5.13, where the distorted and the undistorted signals are plotted vs. time. The assumption is made that at any

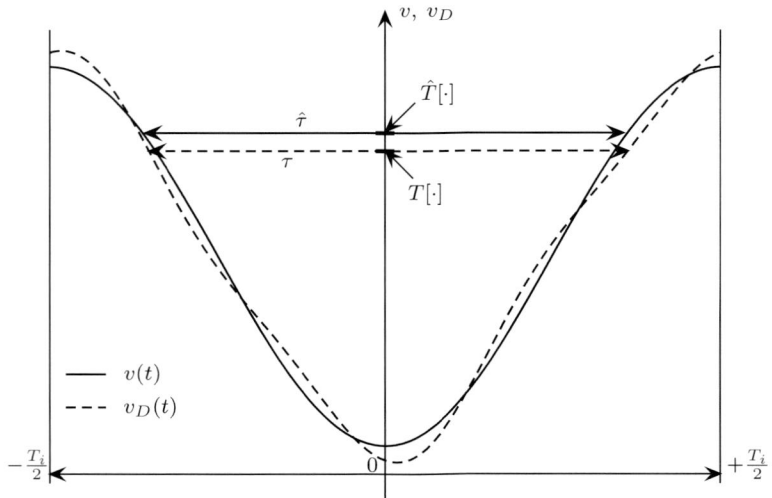

Figure 5.13.   The two intervals, $\hat{\tau}$ and τ, have the same length, and correspond therefore to different voltages, $\hat{T}[\cdot]$ and $T[\cdot]$

point the sign of the derivative of the distorted function is the same as that of the undistorted function. The assumption is questionable near the peaks of the sinewave, which however are ruled out by the presence of overdrive, so that it is not really restrictive, considering typical distortion values. With reference to the non-distorted waveform, probability $\hat{p}$ is represented by the ratio between the duration $\hat{\tau}$ of the interval marked by the thick solid line, and T_i, the sinewave period. The same probability $\hat{p}$, in the case of the distorted function, is represented by the length τ of the interval marked by the dashed line. Let v^{-1} be the inverse function of the non distorted waveform $v(t)$, in the determination corresponding to $[0, T_i/2]$. If $\hat{T}[\cdot]$ is incremented by $d\hat{T}$, interval $\hat{\tau}$ increases by

$$d\hat{\tau} \approx \frac{2\,d\hat{T}}{\left|\frac{dv}{dt}\right|_{t=v^{-1}(\hat{T}[\cdot])}}, \tag{5.77}$$

where the factor 2 accounts for the displacement of the two extremes of interval $\hat{\tau}$. Quite similarly, if the voltage level $T[\cdot]$ is incremented by dT, and v_{Da}^{-1} is the inverse function of $v_D(t)$ in the determination corresponding to $[0, T_i/2]$, v_{Db}^{-1} is the determination corresponding to $[-T_i/2, 0]$,

$$d\tau = dT \left[\frac{1}{\left|\frac{dv_D}{dt}\right|_{t=v_{Da}^{-1}(T[\cdot])}} + \frac{1}{\left|\frac{dv_D}{dt}\right|_{t=v_{Db}^{-1}(T[\cdot])}} \right] \tag{5.78}$$

Thus,

$$\frac{dp}{d\hat{T}[\cdot]} = \frac{1}{T_i}\frac{d\hat{\tau}}{d\hat{T}} \approx \frac{2}{T_i \left|\frac{dv}{dt}\right|_{t=v^{-1}(\hat{T}[\cdot])}}$$

$$\frac{dp}{dT[\cdot]} = \frac{1}{T_i} \left[\frac{1}{\left|\frac{dv_D}{dt}\right|_{t=v_{Da}^{-1}(T[\cdot])}} + \frac{1}{\left|\frac{dv_D}{dt}\right|_{t=v_{Db}^{-1}(T[\cdot])}} \right] \tag{5.79}$$

Considering now that $\frac{dv_D}{dt} = \frac{dv}{dt} - \sum k\omega A_k \sin(k\omega t + \phi_k) \geq \frac{dv}{dt} - \sum k\omega A_k$,

$$\left|\frac{dv_D}{dt}\right|_{t=v_{Da}^{-1}(T[\cdot])} \geq \left|\left[\frac{dv}{dt}\right]_{t=v_{Da}^{-1}(T[\cdot])} - \sum k\omega A_k\right|$$

$$\geq \left|\left|\frac{dv}{dt}\right|_{t=v_{Da}^{-1}(T[\cdot])} - \sum k\omega A_k\right|, \tag{5.80}$$

the last arising from the triangular inequality. A similar inequality holds for determination b.

Now, $v(t)$ is a smooth function, so that $\left|\frac{dv}{dt}\right|_{t=v_{Da}^{-1}(T[\cdot])} \approx \left|\frac{dv}{dt}\right|_{t=v_{Db}^{-1}(T[\cdot])}$, and therefore

$$\frac{dp}{dT[\cdot]} \leq \frac{1}{T_i} \frac{2}{\left|\left|\frac{dv}{dt}\right|_{t=v_{Da}^{-1}(T[\cdot])} - \sum k\omega A_k\right|}. \tag{5.81}$$

The derivatives of the inverse functions are therefore:

$$\frac{d\hat{T}[\cdot]}{dp} = \frac{T_i}{2} \left.\left|\frac{dv}{dt}\right|\right|_{t=v^{-1}(\hat{T}[\cdot])}$$

$$\frac{dT[\cdot]}{dp} \geq \frac{T_i}{2} \left|\left.\left|\frac{dv}{dt}\right|\right|_{t=v_{Da}^{-1}(T[\cdot])} - \sum k\omega A_k\right|, \qquad (5.82)$$

and the following upper bound is determined

$$\left|h'(\hat{p})\right| \leq \frac{T_i}{2} \left\{\left|\left.\left[\frac{dv}{dt}\right]\right|_{t=v^{-1}(\hat{T}[\cdot])}\right| - \left|\left.\left[\frac{dv}{dt}\right]\right|_{t=v_{Da}^{-1}(T[\cdot])}\right| + \sum k\omega A_k\right\}.$$
$$(5.83)$$

Since $v^{-1}(\hat{T}[\cdot])$ is not far from $v_{Da}^{-1}(T[\cdot])$, see Figure 5.13, the two derivatives cancel, so that

$$\left|h'(\hat{p})\right| \leq \sum k\pi A_k. \qquad (5.84)$$

As for $(p_{k+1} - p_k)$, it represents the probability of collecting a sample in code bin $W[k]$, between $T[k]$ and $T[k+1]$. This is always (except possibly when $k = 2^N - 2$) smaller than the probability of collecting a sample with a value between $T[2]$ and $T[1]$

$$p_{k+1} - p_k \leq p_2 - p_1 = P_D\{-A + V_{OD} + W[1]\} - P_D\{-A + V_{OD}\}. \quad (5.85)$$

where V_{OD} is the overdrive and $W[1]$ is the width of the first quantization cell.

If the distortion is small relative to the signal, $P_D\{V\} \approx P\{V\}$ near the peaks of the signal. Therefore

$$p_{k+1} - p_k \leq P\{-A + V_{OD} + W[1]\} - P\{-A + V_{OD}\}. \qquad (5.86)$$

Since the derivative of $P\{V\}$ tends to the infinity as V tends to $-A$, rather than directly attempting a Taylor expansion, it is advisable to consider the inverse function of $P\{V\}$:

$$V \equiv P^{-1}\{P\{V\}\} = -A\cos(\pi P\{V\}) \approx -A + \frac{1}{2}A(\pi P\{V\})^2, \quad (5.87)$$

where the last approximation holds near the negative peak of the sinusoid, for $V \to -A$ and $P\{V\} \to 0$. Solving for $P\{V\}$

$$P\{V\} = \frac{\sqrt{2}}{\pi\sqrt{A}}\sqrt{V + A} \qquad (5.88)$$

and, therefore,

$$P\{-A + V_{OD} + W[1]\} - P\{-A + V_{OD}\} = \frac{\sqrt{2}}{\pi\sqrt{A}}\left(\sqrt{W[1] + V_{OD}} - \sqrt{V_{OD}}\right).$$
$$(5.89)$$

Assuming $W[1] \approx Q$ and substituting (5.84) and (5.89) in (5.75) gives

$$|h(p_{k+1} - h(p_k)| \leq \sqrt{\frac{2Q}{A}} \left(\sqrt{1 + \frac{V_{OD}}{Q}} - \sqrt{\frac{V_{OD}}{Q}} \right) \sum_{i=2}^{h} iA_i. \qquad (5.90)$$

Therefore, if $E_{W dist}$ represents, in code bin widths, the admitted systematic error, it must be

$$\sqrt{\frac{2Q}{A}} \left(\sqrt{1 + \frac{V_{OD}}{Q}} - \sqrt{\frac{V_{OD}}{Q}} \right) \sum_{i=2}^{h} iA_i \leq E_{W dist}Q. \qquad (5.91)$$

For instance, if we assume only second-harmonic distortion and an overdrive $V_{OD} = 1LSB$, to evaluate a code bin width (or a DNL value) for an 8 bit converter with an admitted systematic error of $0.1LSB$, a source distortion smaller than $-36dBc$ is required, which becomes to $-48dBc$ for a 12 bit ADC, for the same overdrive, confidence and admitted error.

5.   Required record length and number of records, expression of measurement uncertainty

This section deals with the estimation of the uncertainty affecting INL (code transition levels) or DNL (code bin width) measurements.

When speaking of uncertainty, it is normally assumed that a correction is applied for each recognized systematic effect that significantly influences the measurement result. One such effect is related to the modification of the input signal p.d.f. due to additive noise: if the noise p.d.f. is known, it is possible, in principle, to calculate the true p.d.f. of the input signal, and by taking the numerical inverse of it, to derive the position of code transition levels.

Quite similarly, it is possible in principle, by spectral analysis, to accurately evaluate the shape of the input waveform, and to compensate the presence of distortion by the use of the corrected p.d.f. and its inverse. This however is highly unpractical, and extremely complex from the computational point of view. Therefore, for the applications, it is preferable to provide an expanded combined uncertainty including systematic errors, and to define an interval where the value of the measurand is believed to lie with a certain confidence.

To this aim, we have computed, in sections 4.1 and 4.2, upper bounds to the errors that may result from applying no correction, bounds which must be properly taken into account in the declaration of measurement uncertainty.

Note, in conclusion, that frequently it is not possible to design the experiment so that the systematic effects are so small, that they can safely be neglected in comparison with random effects, and the test engineer has to face systematic effects comparable to the random ones.

5.1   INL measurements

In order to evaluate the combined standard uncertainty affecting the estimated position of code transition levels, and therefore INL, it is convenient to preliminarily review the relevant contributions .

5.1.1   Systematic contributions.   Let us first consider systematic effects. The systematic error in the position of code transition level $T[k]$ arising from the distortion of the input sinewave has an upper bound expressed by

$$\varepsilon_{dist} \leq \sum_{k=2}^{h} A_k \tag{5.92}$$

where A_k is the amplitude of the kth harmonic and $h = 10$ unless differently specified. We shall assume that, by proper filtering, ε_{dist} has been reduced below a prefixed number $E_{INLdist}$ of code bin widths:

$$|\varepsilon_{dist}| \leq \sum_{k=2}^{h} A_k \leq E_{INLdist}Q \tag{5.93}$$

where $E_{INLdist}$ is the maximum admitted error expressed in code bin widths.

In addition, the modification of the p.d.f. of the input signal resulting from the additive noise introduces a systematic error ε_{pdf} in the position of code transition level $T[k]$ which can be forced below (see section 4.1)

$$\varepsilon_{pdf} \leq E_{INLpdf}Q \tag{5.94}$$

provided that the input overdrive obeys

$$V_{OD} \geq \max\left(2\sigma_n, \frac{0.32\sigma_n^2}{E_{INLpdf}Q}\right). \tag{5.95}$$

Both errors cannot be lowered by averaging.

5.1.2   Random effects.   Taking now care of random effects, additive noise and jitter contribute a variance of the transition levels voltage given by (see (5.31), (5.40))

$$\sigma_{T,n}^2 + \sigma_{T,j}^2 = 1.13\frac{A\pi}{2M}\left(\sigma_n + A\sigma_\phi\right). \tag{5.96}$$

This must be combined with the contribution arising from the intrinsic count uncertainty, taking into account the admitted frequency ratio error (see (5.26))

$$\sigma_{T_{c,intr}}^2 \leq \left(\frac{\pi A}{2M}\right)^2. \tag{5.97}$$

The combined standard uncertainty, expressed in nominal code bin widths $Q = V_{rir}/(2^N - 2) \approx 2^{-N}V_{rir}$, is therefore obtained from

$$u^2_{INL,n+j+c} \leq \frac{1}{2^{-2N}V^2_{rir}}\left(1.13\frac{A\pi}{2M}(\sigma_n + A\sigma_\phi) + \left(\frac{\pi A}{2M}\right)^2\right). \qquad (5.98)$$

In the above expression, the term arising from additive noise can be improved by narrowband filtering; the contribution arising from jitter can (hardly) be reduced by narrowband filtering of the output of the synthesizer providing the clock frequency and by careful optimization of all the jitter sources along the clock chain (waveform shapers, delays,...). All these contributions, in addition, can be effectively reduced by increasing the number of samples M, up to the limit allowed by (5.25) and by the memory depth.

Our objective is to design the experiment, so that the expanded uncertainty associated with the random contributions and corresponding to confidence level $1 - \alpha$ is smaller than an assigned number B of code bin widths. To this aim, it is necessary to determine an appropriate coverage factor k_u: hence, the required combined standard uncertainty, expressed in nominal code bin widths, is readily determined as B/k_u.

The determination of the coverage factor k_u is discussed in section 5.6 under the assumption that the estimated position of a code transition level is a Gaussian random variable with a variance, in code bin widths, expressed by (5.98).

Assuming that the test board has already been optimized, a first question is therefore whether or not the required uncertainty can be achieved with one single record of sufficient length M. To answer the question, consider the inequality

$$\frac{1}{2^{-2N}V^2_{rir}}\left(1.13\frac{A\pi}{2M}(\sigma_n + A\sigma_\phi) + \left(\frac{\pi A}{2M}\right)^2\right) \leq \frac{B^2}{k^2_u}. \qquad (5.99)$$

If it is not possible to find a value of M satisfying conditions (5.99) and (5.25) and compatible with the available memory depth, then it is necessary to acquire several records whose length M is compatible with (5.25).

In fact, if the position estimates obtained from R different records are uncorrelated (to this aim, the assumption was made that the initial phase in a record is a random variable with uniform distribution), the variance is reduced by a factor R. R can therefore be determined from the condition

$$\frac{1}{R\,2^{-2N}V^2_{rir}}\left(1.13\frac{A\pi}{2M}(\sigma_n + A\sigma_\phi) + \left(\frac{\pi A}{2M}\right)^2\right) \leq \left(\frac{B}{k_u}\right)^2 \qquad (5.100)$$

which, by assuming symmetric overdrive and defining the overdrive parameter $c = 1 + 2V_{OD}/V_{rir}$, so that $A = cV_{rir}/2$, becomes

$$R \geq \left(\frac{2^{N-1}k_u}{B}\right)^2 \left(\frac{c\pi}{M}\right) \left(1.13\left(\frac{\sigma_n}{V_{rir}} + \frac{c}{2}\sigma_\phi\right) + \left(\frac{c\pi}{4M}\right)\right). \quad (5.101)$$

It is worth, at this point, to remark that the lowest frequency components of the phase noise associated with the generators play a useful role when multiple records are acquired. In fact, in order to be allowed to apply (5.101), the initial phase in a record should be a random variable with uniform probability density function. To check the validity of this assumption, it is possible to make repeated direct measurements in the time domain, evaluating the phase by least-squares sinewave fitting of the acquired data. This however is very time-consuming.

Alternatively, if the SSB phase-noise spectral density of both generators is known at very low frequency offset from the carrier, the standard deviation of the random shift of the sampling comb with respect to the signal can be roughly estimated as discussed below.

The clock generator phase noise contribution can be evaluated as in (5.37), f_L being the inverse of the delay between the beginning of two subsequent records and f_H, the smallest between f_s/M, the inverse of the record duration, and one half the 3dB bandwidth BW of the selective filter placed at the clock generator output. The signal generator phase noise contribution is evaluated in the same way, considering the bandwidth BW of the filter placed at the ADC signal input.

The two phase noise contributions are then transformed to time-domain and quadratically added: the resulting rms value, under the assumption of normal distribution, should be larger than one half the sampling interval in equivalent time, in order to ensure reasonable record independence. In other terms

$$\sqrt{2\left(\frac{1}{2\pi f_s}\right)^2 \int_{f_L}^{f_{H_{ck}}} \mathcal{L}_{ck}(\zeta)d\zeta + 2\left(\frac{1}{2\pi f_i}\right)^2 \int_{f_L}^{f_{H_{sig}}} \mathcal{L}_{sig}(\zeta)d\zeta} \geq \frac{1}{2Jf_s}.$$
$$(5.102)$$

If, on the contrary, the above inequality is not satisfied, as it sometimes happens using low-noise generators and low M, it may be necessary to deliberately introduce random variations of the initial phase in each record by additional circuitry.

5.1.3   Declaration of uncertainty.

The uncertainty affecting the INL measurement results from both the systematic and the random contributions, and sometimes the systematic contributions cannot be made negligible. In fact, the situation is different for what concerns E_{INLpdf} and $E_{INLdist}$. The former can easily be reduced by increasing the overdrive, with negligible costs. The latter, on the contrary, can only be improved by hardware improvements, and the costs of the improvement may be beyond the planned budget.

So, in what follows, we shall always assume that E_{INLpdf} is so small, that it does not affect the confidence in the results.

One might object, that the converter performance may be degraded by excessive overdrive: this can in any case be solved by avoiding at all to overdrive the input. It is sufficient to limit the transfer characteristic analysis, so that instead of the reduced input range $V_{rir} = T[2^N - 1] - T[1]$ a different range, e.g. $V_* = T[2^N - 10] - T[10]$ is used and declared in the test results.

Making $E_{INLdist}$ negligible, on the contrary, may be an hard job, and the upper bound to distortion shall be carefully accounted for. We know in advance that

$$-E_{INLdist}Q \leq \varepsilon_{dist} \leq E_{INLdist}Q \tag{5.103}$$

and that, centered on ε_{dist}, an interval with half amplitude $B\,Q$ must be allowed to achieve the desired confidence $(1 - \alpha)$.

It is therefore recommended to specify the measurement uncertainty as an interval with half-width

$$Q(E_{INLdist} + B) \quad \text{for safety critical applications;}$$

$$Q\sqrt{\frac{1}{3}E_{INLdist}^2 + B^2} \quad [2] \text{ for less critical applications.} \tag{5.104}$$

where the factor $1/3$ can be justified by the assumption that any value between $-E_{INLdist}$ and $E_{INLdist}$ is equally likely.

It is also strongly recommended to avoid specifying confidence levels higher than 99%, in consideration of the many approximations underlying the developed theory.

5.2   DNL measurements

The sources of uncertainty affecting code bin width or DNL estimates are the same listed in the previous section, but the contributions are different .

5.2.1   Systematic effects.
As before, the systematic effects arising from the modification of the input signal p.d.f. due to additive noise will be neglected, assuming that a sufficient overdrive is used as discussed in section 5.4.1.2. On the contrary, those related to the distortion on the input sinewave have to be considered (see (5.91)).

5.2.2   Random effects.
For what concerns additive noise (section 5.3), if its standard deviation is small in comparison with the code bin width, then the errors in the two transition levels will be almost independent, and the noise contribution to count variance is twice that for code transition levels. If, on the

contrary, the standard deviation of noise is larger than a bin, a strong correlation is expected between the counts corresponding to the two adjacent transition levels which delimit the bin, and the count variance can be sensibly smaller than twice that for a transition level, see (5.36).

For what concerns jitter, the estimates $\hat{T}[k]$ and $\hat{T}[k+1]$ may be treated as uncorrelated random variables. It should be stressed however that, while for additive noise it was shown that correlation reduces variance, so that taking twice the variance for transition levels is always conservative, we did not prove, so far, a similar result for jitter. Thus it is advisable to take into account the possibility of perfect anti-correlation, i.e. taking four times the variance for transition levels, as suggested in (5.105).

For what concerns the intrinsic variance associated with the counting process, it was shown that a sensible correlation exists between the excess counts corresponding to the two adjacent code transition levels. However, the computer simulations reported in [23] adequately support the choice of neglecting such correlation and taking twice the variance (5.26).

In summary, for one single record of length M, the variance affecting the estimated code bin width, in units of code bin widths, is

$$u_{DNL,n+j+c}^2 \leq \frac{2}{2^{-2N}V_{rir}^2}\left(1.13\frac{\pi A}{2M}\left(\sigma^* + 2A\sigma_\phi\right) + \left(\frac{\pi A}{2M}\right)^2\right) \quad (5.105)$$

where $\sigma^* = \min\left\{\sigma_n, \frac{Q}{1.13}\right\}$.

Again, our objective is to design the experiment, so that the expanded uncertainty corresponding to confidence level $(1-\alpha)$ is smaller than B nominal code bin widths. To this aim, it is first necessary to determine the appropriate coverage factor k_u; once this is known, the required combined standard uncertainty expressed in nominal code bin widths is B/k_u.

The coverage factor k_u is determined under the assumption that code bin width estimates have a gaussian distribution with a variance expressed by (5.105).

As before, the first question is whether or not the required uncertainty can be achieved with one single record of sufficient length M. Considering the inequality

$$\frac{2}{2^{-2N}V_{rir}^2}\left(1.13\frac{A\pi}{2M}\left(\sigma^* + 2A\sigma_\phi\right) + \left(\frac{\pi A}{2M}\right)^2\right) \leq \frac{B^2}{k_u^2}, \quad (5.106)$$

if it is not possible to find a value of M satisfying conditions (5.99) and (5.25) and compatible with the available memory depth, then it is necessary to acquire several records.

Considering the width estimates obtained from R different records as uncorrelated, the variance is reduced by a factor R. The appropriate value of R

can therefore be determined from

$$\frac{2}{R\,2^{-2N}V_{rir}^2}\left(1.13\frac{A\pi}{2M}\left(\sigma^* + 2A\sigma_\phi\right) + \left(\frac{\pi A}{2M}\right)^2\right) \le \left(\frac{B}{k_u}\right)^2 \qquad (5.107)$$

and therefore, introducing the overdrive parameter $c = 1 + 2V_{OD}/V_{rir}$, so that $A = cV_{rir}/2$, the condition becomes

$$R \ge 2\left(\frac{2^{N-1}k_u}{B}\right)^2\frac{c\pi}{M}\left(1.13\left(\frac{\sigma^*}{V_{rir}} + c\sigma_\phi\right) + \left(\frac{c\pi}{4M}\right)\right). \qquad (5.108)$$

5.3   Declaration of uncertainty

The declaration of uncertainty in the case of DNL should follow the same guidelines as for INL.

So, the contribution of E_{Wpdf} has to be made negligible by properly choosing the overdrive, as discussed in section 4.1, while E_{Wdist} shall be combined with the extended uncertainty $B\,Q$ either by summing the absolute values or by combining them quadratically, depending on whether the application is safety critical or not.

6.   Choice of the coverage factor

As discussed in the previous section, the position of a code transition level (or the code bin width) estimated from R records may be considered as a random variable whose variance σ_Q^2, expressed in code bin widths, obeys

$$\sigma_Q^2 \le \frac{1}{R\,2^{-2N}V_{rir}^2}\left(1.13\frac{A\pi}{2M}\left(\sigma_n + A\sigma_\phi\right) + \left(\frac{\pi A}{2M}\right)^2\right) \qquad (5.109)$$

for code transition levels, and

$$\sigma_Q^2 \le \frac{2}{R2^{-2N}V_{rir}^2}\left(1.13\frac{A\pi}{2M}\left(\min\left\{\sigma_n, \frac{Q}{1.13}\right\} + 2A\sigma_\phi\right) + \left(\frac{\pi A}{2M}\right)^2\right). \qquad (5.110)$$

for code bin widths.

It is then assumed that the distribution of the estimated code transition levels (respectively, code bin widths) is Gaussian. In fact these estimates result from the sum of many random variables, and in any case the assumption tends to be conservative, because the Gaussian distribution is relatively wide.

The problem is now to determine the appropriate coverage factor k_u in order to ensure the desired extended uncertainty B is not exceeded, whether it is associated with an individual transition level (individual code bin width) or with the worst case of all the code transition levels (all the code bin widths).

Let us first consider the case of an individual transition level. Let B represent the desired extended uncertainty, expressed in nominal code bin widths Q, and $(1 - \alpha)$ the associated confidence level. This means that with probability $(1 - \alpha)$ the position of the transition level should not differ from the estimated value by more than BQ.

Since the displacement of a code transition level from the measured position is a random variable with zero mean and standard deviation $\sigma = Q\sigma_Q$ with normal distribution, then the problem is reduced to the individuation of a suitable coefficient $Z_{0,\alpha/2}$ such that with probability $(1 - \alpha)$ the position of the transition level falls within an interval of amplitude $2Z_{0,\alpha/2}Q\sigma_Q$ centered on the estimated value. Therefore

$$(1 - \alpha) = \frac{1}{\sigma\sqrt{2\pi}} \int_{-Z_{0,\alpha/2}}^{Z_{0,\alpha/2}} e^{-\frac{x^2}{2\sigma^2}} dx = \mathrm{erf}\left(\frac{Z_{0,\alpha/2}}{\sigma\sqrt{2}}\right) = 1 - \mathrm{erfc}\left(\frac{Z_{0,\alpha/2}}{\sigma\sqrt{2}}\right).$$

(5.111)

Thus, $\mathrm{erfc}\left(\frac{Z_{0,\alpha/2}}{\sigma\sqrt{2}}\right) = \alpha$, and taking the inverse

$$Z_{0,\alpha/2} = \sigma\sqrt{2}\,\mathrm{erfc}^{-1}(\alpha). \tag{5.112}$$

Since the confidence level is known, $Z_{0,\alpha/2}$ is readily determined. Then, since the desired extended uncertainty is BQ, it is required that $BQ \geq Z_{0,\alpha/2}Q\sigma_Q$, i.e.

$$\sigma_Q \leq \frac{B}{Z_{0,\alpha/2}} \tag{5.113}$$

The same result holds for an individual code bin width, making reference to the appropriate expression of σ_Q.

Let us now consider an ADC with m code transition levels (respectively, code bins), and let $(1 - \alpha)$ represent the confidence that no level is displaced from its nominal position by more than Z. If the displacements of the individual transition levels are independent random variables, the probability $(1 - \alpha)$ of the event "*no level is displaced from its nominal position by more than Z*" is the product of the probabilities χ of events of the type "*level j is displaced from its nominal position by less than Z*", $j = 1, ..m$. Since the events are assumed as independent, $\chi^m = (1 - \alpha)$, or $\chi = (1 - \alpha)^{\frac{1}{m}}$.

If the displacement of a code transition level from the measured position is a random variable with zero mean and standard deviation σ with normal distribution, then

$$\chi = \frac{1}{\sigma\sqrt{2\pi}} \int_{-Z}^{Z} e^{-\frac{x^2}{2\sigma^2}} dx = \mathrm{erf}\left(\frac{Z}{\sigma\sqrt{2}}\right) = 1 - \mathrm{erfc}\left(\frac{Z}{\sigma\sqrt{2}}\right). \tag{5.114}$$

Hence, $\mathrm{erfc}\left(\frac{Z}{\sigma\sqrt{2}}\right) = 1 - \chi$, and taking the inverse

$$Z = \sigma\sqrt{2}\,\mathrm{erfc}^{-1}(1 - \chi) = \sigma\sqrt{2}\,\mathrm{erfc}^{-1}\left(1 - (1 - \alpha)^{\frac{1}{m}}\right). \tag{5.115}$$

Table 5.2. Values of $Z_{N,\alpha/2}$ for $N = 0, ..., 24$.

v	$Z_{0,\alpha/2}$	$Z_{4,\alpha/2}$	$Z_{8,\alpha/2}$	$Z_{12,\alpha/2}$	$Z_{16,\alpha/2}$	$Z_{20,\alpha/2}$	$Z_{24,\alpha/2}$
.8	1.28	2.46	3.33	4.04	4.64	5.19	5.68
.9	1.64	2.72	3.53	4.21	4.30	5.33	5.81
.95	1.96	2.95	3.72	4.37	4.94	5.46	5.93
.98	2.33	3.22	3.95	4.57	5.12	5.62	6.08
.99	2.58	3.42	4.11	4.71	5.25	5.74	6.19
.995	2.81	3.60	4.27	4.35	5.38	5.85	6.30

For an N bit ADC, $m \simeq 2^N$, so that

$$Z_{N,\alpha/2} = \sigma\sqrt{2}\,\text{erfc}^{-1}\left(1 - (1 - \alpha)^{2^{-N}}\right). \tag{5.116}$$

In summary, when one single code transition level or one single code bin width is of interest, the coverage factor shall be chosen as

$$k_u = Z_{0,\alpha/2} \tag{5.117}$$

while, when the specified confidence refers to the worst case code transition level or code bin width,

$$k_u = Z_{N,\alpha/2}. \tag{5.118}$$

Functions $Z_{0,\alpha/2}$ and $Z_{N,\alpha/2}$ are tabulated in Table 5.2 directly as a function of $v = 1 - \alpha$, i.e. of the desired confidence level.

In accordance with the general recommendations concerning the expression of uncertainty, it is strongly advised to avoid specifying confidence levels larger than 0.99.

7. Comparing the number of samples required by random and by synchronous sampling.

It is interesting to compare the above result with those reported in section 2.2, which refer to the case of perfectly random sampling, with uniform p.d.f. of the sampling phase.

We shall compare INL measurements, considering only the contribution to uncertainty arising from the counting process ($\sigma_n = \sigma_\phi = 0$). In addition, we shall refer to the confidence that the measured value of an individual transition level will not deviate from the expected value by more than BQ, so that $k_u = Z_{0,\alpha/2}$. Thus, (5.19) and (5.101) have to be compared, that is

$$S \geq \frac{Z_{0,\alpha/2}^2 \pi^2 2^{2(N-1)}}{4B^2} \tag{5.119}$$

in the case of random sampling, and

$$S = RM \geq \left(\frac{2^{N-1}Z_{0,\alpha/2}}{B}\right)^2 \left(\frac{c^2\pi^2}{4M}\right) \tag{5.120}$$

in the case of synchronous sampling.

It turns out that the number of samples required in the case of random sampling is larger by a factor M, corresponding to the record length.

This explains the great success of the synchronous sampling approach. In the case of random sampling, it seems in fact that the total number of samples defines the relative uncertainty which can be achieved, independently of the size of the phase interval to be measured. Thus, if a tolerance of a fraction of Q is specified, this is a tight requirement for INL measurements, and so a large number of samples is required. With the synchronous approach, on the contrary, the total number of samples seems to define somehow the absolute uncertainty, whichever is the size of the phase interval to be measured. Thus, less samples are required for the same tolerance.

8.   Determining the transfer characteristic

As discussed in section 5.1, the code transition levels are computed from the cumulative histogram by

$$T[k] = C - A \cos\left(\pi \frac{ch[k-1]}{S}\right) \quad \text{for } k = 1, 2, \ldots, (2^N - 1), \tag{5.121}$$

where A is the amplitude of the sinewave, C is the offset and $S = RM = ch[2^N - 1]$ is the total number of samples.

If the amplitude A and the offset C of the input sinewave are unknown, they can be determined from the data, provided that the position of any two transition levels can be measured by independent means. For instance, if $T[1]$ and $T[2^N - 1]$ are known, A and C may be estimated, respectively, as

$$A = \frac{T[2^N - 1] - T[1]}{\cos\left(\pi \frac{ch[0]}{S}\right) + \cos\left(\pi \left(1 - \frac{ch[2^N-2]}{S}\right)\right)}, \tag{5.122}$$

and

$$C = \frac{T[2^N - 1] \cos\left(\pi \frac{ch[0]}{S}\right) + T[1] \cos\left(\pi \left(1 - \frac{ch[2^N-2]}{S}\right)\right)}{\cos\left(\pi \frac{ch[0]}{S}\right) + \cos\left(\pi \left(1 - \frac{ch[2^N-2]}{S}\right)\right)}. \tag{5.123}$$

An uncorrect estimate of A and/or C will not induce any errors in the estimate of differential or integral nonlinearity, it will only induce gain and offset errors in the transition levels. The transition levels, $T^*[k]$, calculated using the uncorrect estimates $A^* = A + \delta A$ and $C^* = C + \delta C$ will be related to the transition levels, $T[k]$, which would be calculated using the correct values A and C, by the relation

$$T^*[k] = \left(1 + \frac{\delta A}{A}\right) T[k] + C \left(\frac{\delta C}{C} - \frac{\delta A}{A}\right), \qquad (5.124)$$

In high speed dynamic tests, accurate estimates of the amplitude and offset of the sinewave effectively stimulating the converter are difficult to obtain, and it is convenient to refer to the normalized transition levels

$$\overline{T}[k] = -\cos\left(\frac{\pi ch[k-1]}{S}\right) = \frac{T[k] - C}{A}, \qquad (5.125)$$

which are related only to the contents of the cumulative histogram.

9.   Offset error and gain

In a low-frequency test environment, when the amplitude and offset of the input sinewave can be measured directly, the position of code transition levels can normally be determined to within an uncertainty well below one LSB . In such case, *gain* and *offset error* are the values by which the code transition levels, determined using (5.121), are multiplied and then to which the rescaled transition levels are added, respectively,

- to cause the deviation from the nominal transition levels to be zero at the terminal points *(end-points definition)*,

- to minimize the mean squared deviation from the nominal transition levels *(least squares fit definition)*,

- to minimize the maximum of the absolute value of the deviation from the nominal transition levels *(min-max definition)*.

According to these definitions, the transfer characteristic may be represented by

$$G\,T[k] + V_{os} + \epsilon[k] = (k-1)\,Q + T_{nom}[1] = T_{nom}[k], \qquad (5.126)$$

where
$T[k]$ = transition level between codes k and $k-1$, measured by (5.121),
$T_{nom}[k]$ = nominal value corresponding to $T[k]$,
V_{os} = offset error in units of the input quantity (nominally zero),
G = gain (nominally unity),
Q = nominal code bin width,
$\epsilon[k]$ = residual error corresponding to the kth code transition,

and the expression on the right side of (5.126) gives the nominal code transition level, in input units, as a function of k.

At the highest frequencies, since the position of the transition levels cannot be directly measured, it is convenient to transform (5.126) as follows

$$\overline{G}\,\overline{T}[k] + \overline{V}_{os} + \epsilon[k] = T_{nom}[k], \qquad (5.127)$$

so as to refer to the normalized transition levels $\overline{T}$, which are directly determined from the cumulated code histogram. Here, $\overline{G} \equiv GA$ and $\overline{V}_{os} \equiv V_{os} + GC$ are conventional gain and offset parameters, which are used only for the evaluation of DNL and INL. So, while at low frequency the offset and gain errors are useful specification parameters, $\overline{G}$ and $\overline{V}_{os}$ do not bring useful information.

Note that (5.126) and (5.127) are formally identical, and therefore the same formulae will provide (G, V_{os}) and $(\overline{G}, \overline{V}_{os})$ as functions of $T[k]$ and $\overline{T}[k]$, respectively. The equations reported below in sections 9.1, 9.2 and 9.3, which express G and V_{os} as functions of $T[k]$ can therefore be used to evaluate $\overline{G}$ and $\overline{V}_{os}$ by replacing $T[k]$ with $\overline{T}[k]$.

9.1   Gain and Offset (least squares fit definition)

Least squares fit offset and gain are defined as the values of V_{os} and G that minimize the mean squared value of $\epsilon[k]$ over all k . Then, by straightforward calculations,

$$G = \frac{Q\,(2^N - 1)\left(\displaystyle\sum_{k=1}^{2^N-1} k\,T[k] - 2^{(N-1)}\sum_{k=1}^{2^N-1} T[k]\right)}{(2^N - 1)\displaystyle\sum_{k=1}^{2^N-1} T^2[k] - \left(\sum_{k=1}^{2^N-1} T[k]\right)^2} \qquad (5.128)$$

and

$$V_{os} = T_{nom}[1] + Q\left(2^{(N-1)} - 1\right) - \frac{G}{(2^N - 1)}\sum_{k=1}^{2^N-1} T[k]. \qquad (5.129)$$

The values $\epsilon[k]$ which are obtained using the above values of G and V_{os} represent the *least squares fit* integral nonlinearity (see 5.10.1).

9.2   Gain and Offset (end-points definition)

End-points gain and offset are the values that cause $\epsilon[1] = 0$ and $\epsilon[2^N - 1] = 0$, where $2^N - 1$ is the highest code defined . So,

$$G = \frac{T_{nom}[2^N - 1] - T_{nom}[1]}{T[2^N - 1] - T[1]}, \tag{5.130}$$

$$V_{os} = \frac{T_{nom}[2^N - 1]\, T[1] - T_{nom}[1]\, T[2^N - 1]}{T[1] - T[2^N - 1]}. \tag{5.131}$$

Given these values for G and V_{os}, $\epsilon[k]$ is the *end-points* integral nonlinearity (see 5.10.1).

9.3   Gain and Offset (min-max definition)

Min-max gain and offset are the values of G and V_{os} that minimize the maximum of the absolute values $|\epsilon[k]|$, for $k \in [1, 2^N - 1]$. An example of iterative solution is provided below in pseudo code .

```
tol = 1e-9

FOR k=1:(N_code-1)
        sum_T  = sum_T + T[k]
        sum_kT = sum_kT + k * T[k]
        sum_T2 = sum_T2 + T[k] * T[k]
ENDFOR

G = (N_code-1)*(sum_kT - (N_code/2)*sum_T) / ((N_code-1)*sum_T2 - sum_T*sum_T)
Vos = Tn[1] + (N_code/2 - 1) - G*sum_T / (N_code-1)

FOR k=1:(N_code-1)
        sum_err = sum_err + (Tn[k] - Vos - G*T[k])*(Tn[k] - Vos - G*T[k])
ENDFOR

sigma_G = sqrt((N_code-1)*sum_err / ((N_code-3)*((N_code-1)*sum_T2 - sum_T*sum_T)))

inc_G = 3*sigma_G

FIND k: Tn[k] - G * T[k] = max{Tn[] - G * T[]}
FIND j: Tn[j] - G * T[j] = min{Tn[] - G * T[]}
delta = Tn[k] - G * T[k] - Tn[j] + G * T[j]

WHILE (inc_G > tol)
        G_L = G - inc_G
        FIND k: Tn[k] - G_L*T[k] = max{Tn[] - G_L*T[]}
        FIND j: Tn[j] - G_L*T[j] = min{Tn[] - G_L*T[]}
        delta_L = Tn[k] - G_L*T[k] - Tn[j] + G_L*T[j]

        G_R = G + inc_G
        FIND k: Tn[k] - G_R*T[k] = max{Tn[] - G_R*T[]}
        FIND j: Tn[j] - G_R*T[j] = min{Tn[] - G_R*T[]}
        delta_R = Tn[k] - G_R*T[k] - Tn[j] + G_R*T[j]

        IF ((delta_L < delta) AND (delta_L < delta_R))
                G = G_L
                delta = delta_L
        ELSE
        IF ((delta_R < delta) AND (delta_R < delta_L))
                G = G_R
                delta = delta_R
        ENDIF
        inc_G = inc_G / 2
ENDWHILE

FIND k: Tn[k] - G*T[k] = max{Tn[] - G*T[]}
FIND j: Tn[j] - G*T[j] = min{Tn[] - G*T[]}

Vos = (Tn[k] - G*T[k] + Tn[j] - G*T[j]) / 2
```

where Tn is the array containing the nominal code transition levels and N_code is the number of codes (N_code=2^N).

10.   Linearity errors

10.1   Integral nonlinearity

The integral nonlinearity (INL) is the difference $\epsilon[k]$ between the ideal code transition levels $T_{nom}[k]$ and the measured ones, after correcting for gain and offset, $GT[k] + V_{os}$. Integral nonlinearity is usually expressed in LSB or as a percentage of full-scale . The INL definition depends on how gain and offset are defined (least squares, end points or min-max).

When the integral nonlinearity is given as one number without specifying to which code bin it refers, it is intended as the maximum of the absolute value of integral nonlinearity over the entire range.

Once offset error and gain are evaluated in dynamic conditions, using the preferred definition (least squares fit, end-points or min-max), the integral nonlinearity in percent of full scale is

$$INL[k]_\% = -100\%\frac{\epsilon[k]}{V_{fs}} \qquad (5.132)$$

whereas the $INL[k]$ in LSB is

$$INL[k]_{LSB} = -\frac{\epsilon[k]}{Q}. \qquad (5.133)$$

where ϵ is obtained either from (5.126) or from (5.127), depending on whether an independent estimate of A and C is available (low frequency testing) or not (high frequency testing).

When specifying a value of INL, the definition chosen (end-points, least squares fit or min-max) should obviously be indicated.

10.2   Differential nonlinearity

Differential nonlinearity ($DNL[k]$) is the difference, after correcting for gain, between the width of the specified code bin k and the nominal code bin width, divided by the nominal code bin width . When given as one number without code bin specification, it is the maximum of the absolute value of differential nonlinearity over the entire range.

Once the gain G has been determined, as described in section 9, using the appropriate definition, the differential nonlinearity is given by:

$$DNL[k] = \frac{G\left(T[k+1] - T[k]\right)}{Q} - 1. \qquad (5.134)$$

Neither $DNL[2^N - 1]$ nor $DNL[0]$ are defined. Perfect linearity is the same as $DNL = 0$.

When specifying a value of DNL, the DNL definition chosen (end-points, least squares fit or min-max) must be indicated.

In case of high frequency measurements, $DNL[k]$ is evaluated by replacing G with $\overline{G}$ and $T[k]$ with $\overline{T}[k]$ in (5.134).

From the above definition, it is apparent that a **missing code** should be defined as a code k for which

$$DNL[k] = -1. \qquad (5.135)$$

In practice, it is usual to consider as missing a code such that the corresponding DNL is smaller than an assigned value, e.g. -0.80.

11.   Appendix

The results of section 2.4 can be derived by number theory [90], a field of mathematics whose relevance in engineering disciplines is steadily growing.

M samples of the input sinewave are acquired at a constant pace $T_s = 1/f_s$. The frequency $f_i = 1/T_i$ of the input sinewave is related to f_s by $\frac{f_i}{f_s} = \frac{J}{M} + \Delta r$, where J and M are chosen as mutually prime integers, in order to avoid repeated sampling of the input sinewave at the same phase, and Δr is the frequency ratio error.

Taking the origin of the time axis in correspondence of the first sample, with index 0, the sampling instants are: $0, T_s, 2T_s, ...(M-1)T_s$. Let us represent by $\lfloor x \rfloor$ the largest integer $\leq x$ and by $\langle x \rangle$ the difference between x and $\lfloor x \rfloor$. When the samples are rearranged in the phase domain, where the period T_i of the input sinewave corresponds to 2π, the phase of the $n-th$ sample is represented by 2π times the quantity $\left\langle \frac{nT_s}{T_i} \right\rangle$, Note that $\left\langle \frac{nT_s}{T_i} \right\rangle = \left\langle n\left(\frac{J}{M} + \Delta r\right) \right\rangle$. Neglecting the factor 2π one is led to consider a vector $\mathbf{x} \overset{\triangle}{=} [0 \ \ x_1(\Delta r)...x_{M-1}(\Delta r)]$ which defines the sampling phases normalised to 2π, where

$$x_n(\Delta r) \overset{\triangle}{=} \left\langle n\left(\frac{J}{M} + \Delta r\right) \right\rangle, \quad n = 0, ..., M-1. \qquad (5.136)$$

Note that, with this definition, the elements of the vector are arranged in the same order as in the acquired data record, and not in order of increasing equivalent phase. This can be better appreciated by arranging the sampling points, marked by dots, along the trigonometric circle, as shown in Figure 5.14, assuming as usual that phase increases in the counterclockwise direction. The example in the figure corresponds to $M = 7, J = 2, \Delta r = 0$.

A useful lemma

We shall first show that

if $|\Delta r| < 1/[M(M-1)]$, the magnitude of the distance in phase between each sample and the position that the same sample would have for $\Delta r = 0$ is

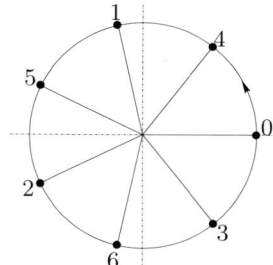

Figure 5.14. Each sampling phase, marked by a dot, is labelled by a number which indicates the position in the input record. The example corresponds to perfectly coherent sampling with $M = 7, J = 2$.

bounded by $2\pi/M$, i.e.

$$|x_n\,(\Delta r) - x_n(0)| < \frac{1}{M} \tag{5.137}$$

Proof:

From (5.136)

$$|x_n\,(\Delta r) - x_n(0)| \triangleq \left|\left\langle n\left(\frac{J}{M} + \Delta r\right)\right\rangle - \left\langle n\frac{J}{M}\right\rangle\right| \quad n = 0, ..., M - 1 \tag{5.138}$$

Since $\langle x \rangle \triangleq x - \lfloor x \rfloor$, (5.138) becomes

$$|x_n\,(\Delta r) - x_n(0)| \triangleq \left| n\Delta r - \left\lfloor n\frac{J}{M} + n\Delta r \right\rfloor + \left\lfloor n\frac{J}{M} \right\rfloor \right| \tag{5.139}$$

Since $|\Delta r| < [M(M-1)]^{-1}$, it follows

$$n|\Delta r| \leq (M-1)|\Delta r| < \frac{1}{M} \quad n = 0, 1, ..., M - 1 \tag{5.140}$$

As a consequence,

$$\left\lfloor n\frac{J}{M} + n\Delta r \right\rfloor = \left\lfloor n\frac{J}{M} \right\rfloor \tag{5.141}$$

and therefore

$$|x_n\,(\Delta r) - x_n(0)| = |n\Delta r| < \frac{1}{M}. \tag{5.142}$$

Measuring distances between sample points in the equivalent phase domain

Figure 5.15 shows how the sampling phases are arranged in a typical case of quasi-synchronous sampling with $M = 7, J = 2$ and $\Delta r = 1/196$. It may

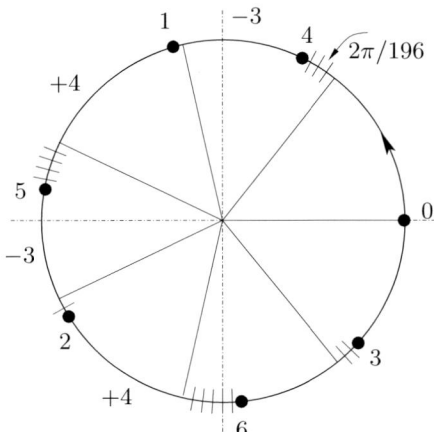

Figure 5.15. Each sampling phase, marked by a dot, is labelled by a number which indicates the position in the input record. The example corresponds to quasi-synchronous sampling with $M = 7, J = 2$ and $\Delta r = 1/196$. The thin dashes mark the distance of the sampling point from the position corresponding to $\Delta r = 0$ in units of $2\pi/196$.

be noted that, when the samples are arranged in order of increasing equivalent phase, the following sequence results: $x_0, x_4, x_1, x_5, x_2, x_6, x_3$. The same sequence would be found for $\Delta r = 0$.

A first relevant question is: given a sample in the sequence, say x_4, how can I find the index of, say, the second nearest sample in the direction of increasing equivalent phase? The obvious solution is, in this case, 5, i.e. x_5 is the required sampling phase.

The problem may be analysed with the help of Farey series [70]. The Farey series of order M, $\mathcal{F}(M)$ is defined as the set of all fractions in lowest terms between 0 and 1 whose denominators do not exceed M, arranged in increasing order. As an example, $\mathcal{F}(7)$ is represented by the sequence

$$\frac{1}{7} \ \frac{1}{6} \ \frac{1}{5} \ \frac{1}{4} \ \frac{2}{7} \ \frac{1}{3} \ \frac{2}{5} \ \frac{3}{7} \ \frac{1}{2} \ \frac{4}{7} \ \frac{3}{5} \ \frac{2}{3} \ \frac{5}{7} \ \frac{3}{4} \ \frac{4}{5} \ \frac{5}{6} \ \frac{6}{7} \ . \tag{5.143}$$

One has to look, in the Farey series of order M, for the two terms which surround J/M: in our case, where $J = 2$ and $M = 7$,

$$\frac{1}{4} < \frac{2}{7} < \frac{1}{3} \tag{5.144}$$

In general, these two terms will be represented as follows

$$\frac{J_L}{M_L} < \frac{J}{M} < \frac{J_R}{M_R}, \tag{5.145}$$

relationship which defines J_L, M_L, J_R, M_R. Coming back to our problem, given a reference sampling phase x_n, let us define the distance between the

$K - th$ nearest sample in the direction of increasing phases and x_n as

$$d_{n,K}(\Delta r) \stackrel{\triangle}{=} \langle x_{n+m}(\Delta r) - x_n(\Delta r) \rangle \quad n = 0, 1, ..., M - 1 \qquad (5.146)$$

where the argument Δr stands to remember that this distance depends on the frequency ratio error, and $-M < m < M$ is a suitable integer depending on K. When the error is zero, this distance is obviously K/M. So, the equation

$$d_{n,K}(0) = \frac{K}{M}, \quad n = 1, ..., M - 1 \qquad (5.147)$$

may be used to implicitly define m.

Let us call *"target"* index the index of the sample which nominally should be at the distance $2\pi \frac{K}{M}$ from the reference. If the target index is larger than the reference index, then m must be greater than 0; if the target index is smaller than the reference index, one must choose $m < 0$. It turns out that there are two solutions, m_L and m_R, of (5.147) in the range $-M < m < M$. In the example chosen, the two solutions are $m_L = 1$ and $m_R = -6$, the last being justified by the fact that subtracting 6, modulo 7 is equivalent to adding 1. However, one should consider that $n + m$ must be in the range $0, ..., M - 1$, so m_L can be retained ($4 + 1 = 5$), while m_R has to be discarded ($4 - 6 = -2$).

More generally, by considering the definition of $x_n(\Delta r)$ for $\Delta r = 0$, (5.147) becomes

$$\left\langle \left\langle (n + m) \frac{J}{M} \right\rangle - \left\langle n \frac{J}{M} \right\rangle \right\rangle = \frac{K}{M} \qquad (5.148)$$

which becomes

$$\left\langle m \frac{J}{M} \right\rangle = \frac{K}{M} \qquad (5.149)$$

considering that $\langle \langle x \rangle \pm \langle y \rangle \rangle = \langle x \pm y \rangle$.

By assuming at first $m > 0$, (5.149) becomes

$$mJ - uM = K \qquad (5.150)$$

where u is the largest integer smaller than or equal to $m \frac{J}{M}$. Assuming that $0 < K < M$, since we are dealing with equivalent phases, (5.150) becomes

$$mJ \bmod M = K. \qquad (5.151)$$

It was shown, in [27], that $m = M_L$ is a solution of

$$mJ \bmod M = 1, \qquad (5.152)$$

and therefore $K M_L$ is a solution of (5.151), and $m_L \stackrel{\triangle}{=} (K M_L \bmod M)$ is a solution of (5.147), positive and smaller than M. When looking for a negative solution, a similar proof leads to $m = m_R = -(K M_R \bmod M)$.

Having so determined m_L and m_R, we are now in a position to show that $d_{n,K}(\Delta r)$, the distance of two sampling phases which should be nominally spaced by $2\pi\frac{K}{M}$ can only take one of the two values:

$$d_{L,K}(\Delta r) \triangleq \frac{K}{M} + m_L \Delta r , \quad d_{R,K}(\Delta r) \triangleq \frac{K}{M} + m_R \Delta r . \qquad (5.153)$$

Assume for instance that $\Delta r > 0$, that is the sampling interval is larger than due, and consider again the case of Figure 5.15 ($M = 7, J = 2, \Delta r = 1/196$). The sampling points for $\Delta r = 0$ are represented by the thin radial lines; the actual positions of the sampling points by the labelled dots. For each sampling position, the distance from the position it would occupy in the case $\Delta r = 0$ is marked in units of $2\pi\Delta r$.

Note that the 2π interval is covered in two sweeps, $\{0, 1, 2, 3\}$ and $\{4, 5, 6\}$, in the case $J = 2$; in general, J sweeps are required to cover all the sampling phases. Considering for instance as a reference x_4, it is apparent that the distance between it and the target sample (k^{th} nearest neighbour in the direction of increasing equivalent phase) can be either greater or smaller than $2\pi K/M$, always by integer multiples of $2\pi\Delta r$.

Now, if the target index is larger than the reference index, then $m > 0$, and one must choose $m = m_L$; if the target index is smaller than the reference index, then one must choose $m = m_R < 0$. In any case, the other choice is discarded, because it leads to an index outside the range $0, ..., M - 1$.

Then, assuming a positive Δr, if the target index is larger than the reference index, at each intermediate step moving from reference to target one gains an extra $2\pi\Delta r$, so that, when the target is reached, one has accumulated an extra $2\pi m_L \Delta r$. The situation can be conveniently examined in the above figure, considering sample x_4 as the reference and $K = 4$. Here, $m_L = 4M_L \bmod 7 = 2$, and in fact, moving from x_4 to x_6 one gains $2\Delta r$, so that the overall distance becomes $2\pi(\frac{4}{7} + 2\Delta r)$.

If one considers the same reference sample x_4 and $K = 3$, in order to reach the target x_2 one has to move backwards in the sequence of samples, and as a consequence the target sample is less displaced from the nominal position than the reference sample. Here, $m_R = -(3M_R \bmod 7) = -2$, and the overall distance becomes $2\pi(\frac{3}{7} - 2\Delta r)$.

Now, obviously the sum of the distances of all the nearest neighbours spans 2π, or, in normalized terms,

$$\sum_{n=0}^{M-1} d_{n,1}(\Delta r) = 1. \qquad (5.154)$$

Quite similarly, if one sums the distances of all the second $(K - th)$ nearest neighbours, two (respectively K) rotations around the origin are performed

$$\sum_{n=0}^{M-1} d_{n,K}(\Delta r) = K. \tag{5.155}$$

Thus,

$$\sum_{n=0}^{M-1} d_{n,K}(\Delta r) = h\, d_{L,K}(\Delta r) + k\, d_{R,K}(\Delta r) = K. \tag{5.156}$$

where $d_{L,K}$ and $d_{R,K}$ are defined in (5.153), and h and k, with $h + k = M$, are two suitable integers which represent the number of occurrences of $d_{L,K}$ and $d_{R,K}$, respectively, when n sweeps the range $0, ..., M - 1$. From the property of Farey series [90] it results $M_L + M_R = M$. It may be shown that, as a consequence, $m_R = m_L - M$. From (5.153) and (5.156) it follows

$$h = -m_R, \quad k = m_L. \tag{5.157}$$

Derivation of the count probabilities

In this section the probabilities associated with the possible contents of $ch[k - 1]$ are derived, and hence the count variances reported in Table 5.1.

Let $v(t) = A\cos(\omega t + \phi) + C$ represent the input signal. For assigned values of ψ_k, J, M and Δr, $ch[k - 1]$ is a deterministic function of ϕ, which can be determined by counting the number of phase samples that fall inside the angle $(-\psi_k, \psi_k)$ as a function of ϕ. By normalizing phase angles to 2π, this is the same as the number of samples inside the sector $\phi/(2\pi)$, $(\phi + 2\psi_k)/(2\pi)$, with $0 \leq \phi/(2\pi) < 1$.

The situation is represented in Figure 5.16 for the case of perfectly synchronous sampling with $M = 7, J = 2$. The gray sector represents the normalised angle $2\psi_k/2\pi = n_k + \alpha_k$, where $n_k = 3$ and $\alpha_k = 0.09$, and its position is defined by ϕ. In the outer ring, the number of counts in excess of n_k is represented by different levels of gray as a function of ϕ. As ϕ increases, $ch[k - 1]$ is equal either to n_k or to $n_k + 1$. For instance, for $0 < \phi/(2\pi) < (1 - \alpha_k)/M$ the number of counts is n_k, and becomes $n_k + 1$ for $(1 - \alpha_k)/M < \phi/(2\pi) < 1/M$. In a complete rotation, the same pattern is repeated seven times, so that, if ϕ is uniformly distributed in $[0, 2\pi)$, he probability of counting n_k is $1 - \alpha_k$, while the probability of counting $n_k + 1$ is α_k. When $\Delta r \neq 0$, the same approach may be adopted. Let us consider first the case $\Delta r > 0$. Figure 5.17 refers to the case $J = 2, M = 7, n_k = 3, \alpha_k = 0.09$ and $\Delta r = 1/196$, and again the levels of gray in the outer ring indicate the counts in excess of n_k. As may be noted, the sampling phases

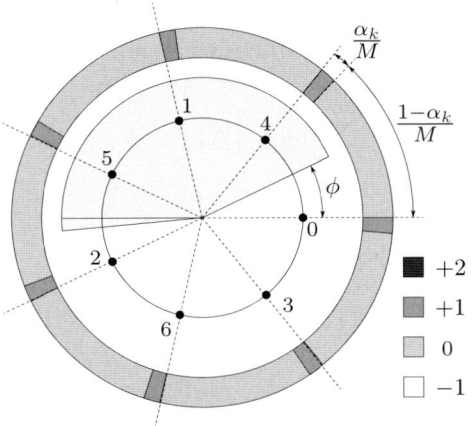

Figure 5.16. Coherent sampling. Number of counts as a function of sinewave phase, assuming $J/M = 2/7, n_k = 3, \alpha_k = 0.09$ and $\Delta r = 0$. Levels of gray in the outer circle indicate counts in excess of n_k.

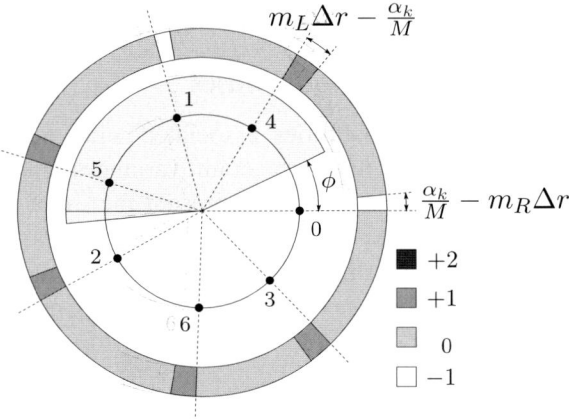

Figure 5.17. Noncoherent sampling. Number of counts as a function of sinewave initial phase, assuming $J/M = 2/7, n_k = 3, \alpha_k = 0.09$ and $\Delta r = 1/(4M^2)$. Levels of gray in the outer circle indicate counts in excess of n_k.

are no longer equally spaced. In any case, if the condition for the validity of (5.137) is satisfied, $|d_{L,K}(\Delta r) - d_{L,K}(0)| = |m_L \Delta r| < 1/M$, and similarly $|d_{R,K}(\Delta r) - d_{R,K}(0)| \leq 1/M$. From this, it follows that $ch[k-1]$ may only assume values in the interval $[n_k - 1, n_k + 2]$. More precisely, the only possible counts are: $n_k - 1, n_k, n_k + 1$ or $n_k, n_k + 1$ or $n_k, n_k + 1, n_k + 2$, depending on the value of α_k. Part a) of Figure 5.18 demonstrates the possibility of counting $n_k - 1$ with reference to the example $J = 2, M = 7, n_k = 3, \alpha_k = 0.09$ and $\Delta r = 1/196$. This occurs as long as $0 \leq \phi \leq \vartheta_1$. Similarly, part b) of the figure demonstrates the possibility of counting $n_k + 2 = 5$ for the same values of the parameters, except $\alpha_k = 0.91$. Five counts are obtained when Φ falls in the phase interval marked by ϑ_3. From (5.153) it results that $d_{L,n_k}(\Delta r) > n_k/M$

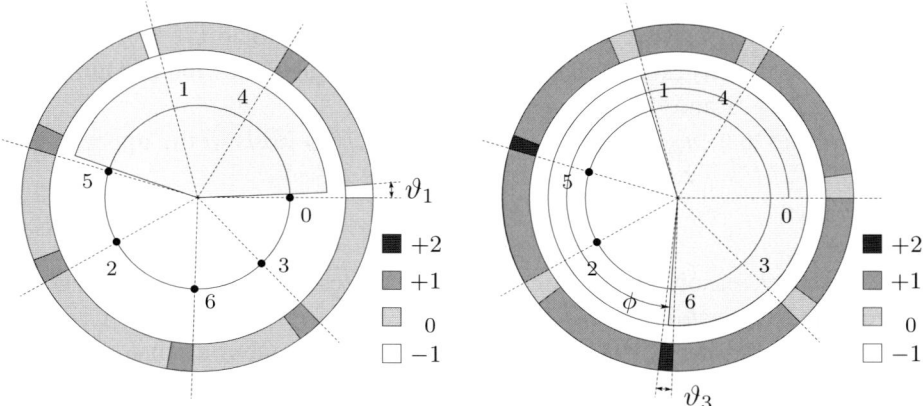

Figure 5.18. a): a condition where $n_k - 1$ is counted. Choosing $J = 2$, $M = 7$, $n_k = 3$, $\alpha_k = 0.09$ and $\Delta r = 1/(4M^2)$, for positions of the gray sector representing $2\psi_k$ between the two shown, 2 counts are collected. b): a condition where $n_k + 2$ is counted. Now $\alpha_k = 0.91$, all the other parameters remaining equal. For positions of the gray sector representing $2\psi_k$ between the two shown, 5 counts are collected.

and $d_{R,n_k}(\Delta r) < n_k/M$. It follows that $ch[k-1] = n_k - 1$ when

$$\frac{2\psi_k}{2\pi} = \frac{n_k + \alpha_k}{M} < d_{L,n_k}(\Delta r) = \frac{n_k}{M} + m_L\Delta r \qquad (5.158)$$

i.e. when $\alpha_k < Mm_L\Delta r$. From (5.157), d_{L,n_k} occurs $-m_R$ times as n increases form 0 to $M - 1$. For each of these intervals of width d_{L,n_k} reference and target sampling phases are defined. Consider that $n_k - 1$ is counted, as the sector of width $2\psi_k$ rotates counterclockwise, from the moment when the lagging edge of the rotating sector coincides with the reference sampling phase until the moment when the leading edge of the rotating sector reaches the target sampling phase, i.e. for a phase interval of width $\frac{n_k}{M} + m_L\Delta r - \left(\frac{n_k + \alpha_k}{M}\right) = m_L\Delta r - \alpha_k/M$. Thus, the probability of the event " $n_k - 1$ *counts* " can be written as

$$p_{-1} = \left(\frac{\alpha_k}{M} - m_L\Delta r\right) m_R. \qquad (5.159)$$

Quite similarly. denoting by n_L and n_R the two solutions of (5.24), the event $ch[k-1] = n_k + 2$ may occur if

$$\frac{2\psi_k}{2\pi} = \frac{n_k + \alpha_k}{M} > d_{R,n_k+1}(\Delta r) = \frac{n_k + 1}{M} + n_R\Delta r \qquad (5.160)$$

i.e. when $\alpha_k > 1 + Mn_R\Delta r$.

When $Mm_L\Delta r < \alpha_k < 1 + Mn_R\Delta r$, $ch[k-1]$ equals either n_k or $n_k + 1$, as it occurs for perfectly coherent sampling.

If

$$Mm_L\Delta r < 1 + Mn_R\Delta r \tag{5.161}$$

it is not possible for an α_k to be simultaneously greater than $1 + Mm_L\Delta r$ and smaller than $Mm_L\Delta r$, so that condition (5.161) precludes the possibility of counting $n_k - 1$ and $n_k + 2$ for the same α_k.

Recalling that $|n_R|, m_L < M$, condition (5.161) leads to the upper bound $\Delta r < 1/(2M^2)$, i.e.

$$\frac{\Delta r}{r} < \frac{1}{2JM} \tag{5.162}$$

By similar reasoning the other entries in Tables 5.3 and 5.4 may be determined, and since $ch[k-1]$ can only take on values in the range $[n_k - 1, n_k + 2]$ p_0 may be determined from the other entries by remembering that $p_0 = 1 - p_{-1} - p_1 - p_2$.

Table 5.3.   Probabilities of the number of counts for $0 \le \Delta r \le 1/(2M^2)$.

	$0 \le \alpha_k < Mm_L\Delta r$	$Mm_L\Delta r \le \alpha_k < 1 + Mn_R\Delta r$	$1 + Mn_R\Delta r \le \alpha_k < 1$
p_{-1}	$\left(\frac{\alpha_k}{M} - m_L\Delta r\right)m_R$	0	0
p_1	$\left(\frac{\alpha_k}{M} - m_R\Delta r\right)m_L$	α_k	$\alpha_k - 2p_2$
p_2	0	0	$\left(\frac{\alpha_k}{M} - \frac{1}{M} - n_R\Delta r\right)n_L$

Table 5.4.   Probabilities of the number of counts for $-1/(2M^2) < \Delta r < 0$.

	$0 \le \alpha_k < Mm_R\Delta r$	$Mm_R\Delta r \le \alpha_k < 1 + Mn_L\Delta r$	$1 + Mn_L\Delta r \le \alpha_k < 1$
p_{-1}	$\left(m_R\Delta r - \frac{\alpha_k}{M}\right)m_L$	0	0
p_1	$\left(m_L\Delta r - \frac{\alpha_k}{M}\right)m_R$	α_k	$\alpha_k - 2p_2$
p_2	0	0	$\left(\frac{1}{M} - \frac{\alpha_k}{M} + n_L\Delta r\right)n_R$

From these, it follows that the expected value of $ch[k]$ is $n_k + \alpha_k$, as anticipated in (5.22) and the contents of Table 5.1 may be derived by considering that

$$\sigma_{c,intr}^2 = (1 - 2n_k)p_{-1} + (1 + 2n_k)p_1 + 4(1 + n_k)p_2 - \alpha_k(\alpha_k + 2n_k). \tag{5.163}$$

Chapter 6

COMPARATIVE STUDY OF ADC SINEWAVE TEST METHODS

José Machado da Silva
Universidade do Porto, FEUP – INESC-Porto
Campus da FEUP, Rua Dr Roberto Frias
4200-465 Porto, Portugal
jms@fe.up.pt

Hélio Mendonça
Universidade do Porto, FEUP – INESC-Porto
Campus da FEUP, Rua Dr Roberto Frias
4200-465 Porto, Portugal
hsm@fe.up.pt

Sara Mazoleni
ITALTEL SpA, Procurement Department - Supplier Quality
Via Reiss Romoli
I-20019 Castelletto di Settimo Milanese
Milan, Italy
Sara.Mazzoleni@italtel.it

1.   Introduction

Several issues must be addressed when selecting a test method. Whether performing tests in the laboratory or at the production stage, the first concern is efficiency, that is, the capacity of providing reliable results (in fact estimates) for the parameters being measured in the laboratory, and the capacity of accepting good parts and rejecting the bad ones without errors in production, in both cases over a large number of repeated operations. Accuracy and precision are thus the two first key aspects to be evaluated. They are directly affected by the test setup, as well as, by the data processing algorithms.

D. Dallet and J. Machado da Silva, (eds.), Dynamic Characterisation of
Analogue-to-Digital Converters, 157–215.

The most meaningful way of specifying the accuracy of a parameter estimate $\hat{P}$ is to determine confidence intervals $\hat{P}^l \leq \hat{P} \leq \hat{P}^u$, found from measurement results, where the probability of finding the expected values of the estimated parameters is high — $prob[\hat{P}^l \leq \hat{P} \leq \hat{P}^u] \geq (1 - \epsilon)$ where $(1 - \epsilon)$ is the confidence level, and ϵ is small.

Test equipment cost is another very important issue. Although accuracy and precision are directly related to and determined by the quality/cost ratio of the equipment and of the overall performance of the test setup being used, it is important to evaluate whether a certain test or measure can be performed with the same accuracy and precision using a different method, less demanding in terms of the test setup performance quality. In the ADC testing domain this is also determined by the quality of test stimulus and clock signals (distortion and noise). If these do not satisfy the minimum required levels, filters must be used to reduce harmonic distortion and phase variations. In the case of high resolution ADCs it is particularly necessary to reduce stimulus noise and harmonic distortion using bandpass filters. Suitable noise reduction mechanisms are also required on the clock path in order to reduce jitter between test stimulus and clock signals.

Testing time is normally not an issue in laboratory tests, but in production it is actually a critical one. Due to ATE (Automatic Test Equipment) high costs, fast tests are crucial to increase the number of parts tested per unit of time and consequently to reduce test equipment cost of ownership. In the ADC testing process, testing time is determined by both data acquisition (number of records × record duration) and data processing times, which are dependent on the number of samples in the record (M) and on the complexity of the algorithms used to process them. Relay switching and filter settling times are other delays which affect data acquisition time.

According to these considerations, the accuracy and precision of the three ADC test methods are evaluated here in terms of their sensitivity to the following parameters

- number of captured samples

- stimulus distortion and noise

- stimulus offset

- sampling clock jitter

- coherence of the sampling process

The methods were also evaluated in terms of testing time, by comparing the number of cycles and samples required to obtain a certain accuracy and precision, and the computation time required by the respective data processing algorithm.

2. General considerations

2.1 Errors introduced by the test setup

The block diagram of a typical test setup for ADC dynamic testing can be seen in figure 6.1. The input stimulus of the ADC-UT (ADC under test) is a sinewave, which is well known from the mathematical point of view and is easily obtainable with a reasonable purity using synthesizers and filters available in the market.

Nonlinear effects generally increase out of proportion with the signal amplitude, so that some failures will only be seen with a full-scale input. The test sine waveform has a nonuniform probability density function in the amplitude domain because it presents a time variable slope. This leads the ADC-UT transition levels to be stimulated in a different way along the transfer function characteristic. A transition level displacement with respect to the ideal value (a nonlinearity error) leads to an increase in distortion and quantisation noise powers, which are dependent on the displacement location on the transfer characteristic.

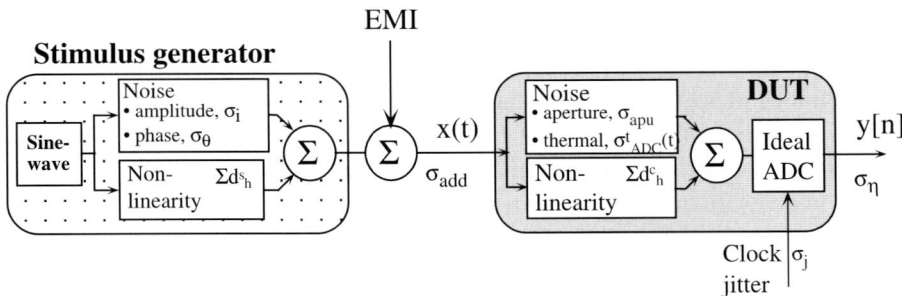

Figure 6.1. Sources of noise and distortion in an ADC test setup.

These aspects affect the selection of the test stimulus frequency. Testing at a single frequency might not provide an evaluation of how the ADC behaves in the bandwidth required for the application. However, this depends on the parameter being evaluated. Since DNL does not change with input frequency, so does not SNR as both parameters are related each other [74]. But, SNR might decrease at higher frequencies due to the increase of higher frequency harmonics which cannot be distinguished from noise floor, and thus may be included in the SNR computation. On the other hand, THD is directly related to the input frequency. Below a certain frequency THD is only affected by INL, but as frequency increases INL worsens, as well as THD, due to mainly the dynamic performance of the sample and hold operation [75].

Performing tests at various frequencies increases significantly both testing time and equipment cost, the latter due to the variety of filters which are eventually necessary.

The test setup is responsible for noise and interferences which can lead to test errors. However, all the tests are intrinsically dependent on the quality of the sinewave source and the overall test setup — hardware, surrounding environment, and quality and characteristics of the instrumentation being used — which can lead to test inaccuracy and non-repeatable results. For example in production testing, the additional circuitry and instrumentation required to perform different tests might degrade noise performance. The cables which connect the stimulus generator to the ADC-UT might introduce parasitics which degrade the measurement quality.

At increasing frequencies the output response of the ADC-UT tends to be actually a combination of device under test and test setup behaviours. Figure 6.1 illustrates the major sources of noise and distortion which are unrelated to the ADC-UT intrinsic behaviour. These are amplitude noise, distortion and phase jitter of the input stimulus, and sampling clock phase noise (clock jitter). Power supply noise or poor voltage regulation is another source of error which directly affects the ADC-UT performance. The signal conditioning circuitry can also introduce nonlinearities due to, for instance, phase delays and impedance mismatch with the input of the ADC.

A number of samples (M) of the observed digital output sequence are acquired through a data acquisition board, and finally processed using suitable dedicated signal processing algorithms, in order to determine the ADC parameters to be evaluated. The only deviations present in the captured data should be noise intrinsic to the ADC-UT functional behaviour, and noise and harmonics caused by the ADC's nonlinearities.

However, the actual SNR of the acquired data is

$$SNR = -10 \log[(2\pi f_{in}\sigma_{jit})^2 + (\frac{\sigma_{in}}{2^N})^2 + (2\pi f_{in}\sigma_{apu})^2 +$$
$$+ (\frac{\sigma_{ADC}^t}{2^N})^2 + \frac{1}{3.2^{2N-1}}] \qquad (6.1)$$

where, σ_{jit} is the relative *rms* jitter of stimulus and clock signals, σ_{in} is the test stimulus *rms* noise, σ_{apu} the standard deviation of the ADC aperture uncertainty, σ_{ADC}^t the ADC intrinsic *rms* noise, and $\frac{1}{3.2^{2N-1}}$ the ideal SNR due to quantisation. The terms $2\pi f_{in}\sigma_{jit}$ and $2\pi f_{in}\sigma_{apu}$ coarsely represent upper bounds to the amplitude uncertainty of the sampled code due to the uncertainty in the sampling instant. They are proportional to the signal frequency and, typically, their maximum occurs for input amplitudes around the average value of the sinewave amplitude, where the slope of the stimulus is maximum.

The distortion generated by the quantisation process presents characteristics similar to those caused by additive noise. This assumption leads to model the quantiser output ($y[n]$) as the sum of an infinite precision input stimulus ($x[n]$), a quantisation error $q[n]$ and an an additive white Gaussian random process $w[n]$ with zero mean and variance σ_{add}^2 uncorrelated with the test input stimulus, i.e., $y[n] = x[n] + q[n] + w[n]$. Furthermore, since $q[n]$ and $w[n]$ can be considered uncorrelated, the overall error (noise) can also be modelled as a white Gaussian process with variance $\sigma_\eta^2 = \sigma_q^2 + \sigma_{add}^2$. A high random noise content in the captured data may require the acquisition of several records of data in order to minimise noise after averaging those records.

As we will see later coherent sampling is generally required, and thus synchronisation between the sinewave synthesizer and the clock generator is necessary. This requirement increases the test setup complexity and restricts the range of test stimulus and sampling clock frequencies.

2.2 Test methods based on spectral analysis and sine fitting

2.2.1 Spectral Analysis using the Fourier Transform.
This technique applies the discrete Fourier transform (DFT) to the captured data record in order to obtain its representation (spectra) in the frequency domain . Besides the main line representing the input fundamental sinewave frequency, several other features present on the spectrum provide additional information about the ADC's performance.

The smallest resolution bandwidth achievable with DFT is

$$\Delta f \;\; = \;\; \frac{ENBW \times f_s}{M} \qquad (6.2)$$

$$ENBW \;\; = \;\; M \times \frac{\sum w(i)^2}{[\sum w(i)]^2} \qquad (6.3)$$

where f_s is the sampling frequency and M the length of the DFT used. ENBW (equivalent noise bandwidth) is a factor to adjust the resolution bandwidth when a window is used, $w(i)$ being the weights of the window.

From the frequency spectrum obtained it is possible to extract a set of parameters that make possible the characterisation of the converter in terms of output distortion and noise. The most important ones are

- SINAD - Signal to noise and distortion ratio

- SNR - Signal to noise ratio

- THD - Total harmonic distortion

- SFDR - Spurious free dynamic range

From these it is also possible to obtain other parameters like the effective number of bits (N_{ef}). All these parameters take into account the harmonic distortion and/or the noise present on the captured data, and thus are prone of being directly affected by the behaviour of the test setup. In fact, one of the drawbacks associated with the SA method is the non-uniformity of the noise floor of a single data record transform. As the noise floor represents a set of reasonably well behaved random data it can be smoothed by averaging several spectra (tens or even hundreds) obtained from successive data collections, in order to limit the noise floor to a level suitable for measurement. By averaging the power spectrum of R different data records the variance of the ADC noise in Δf is reduced by R. This might be computationally intensive and time consuming compared to the data collection time.

Besides noise, spectral ADC test parameters are also sensitive to offset and amplitude deviations of the sinusoidal test stimulus. It has been shown that a deviation of less than one tenth of one nominal quantisation level (Q) can lead to a deviation of 10dB or more in the measured THD. This effect outcomes from the rounding operation that takes place in the quantisation process, which affects particularly the low-frequency components [45]. Offset affects also the variance of the quantisation noise [73].

Spectral analysis yields an overall performance behaviour of the ADC-UT under the specified test conditions. Due to the discontinuous nature of the ADC transfer characteristic it is not immediate to extrapolate the ADC response for different input signals. Further tests have to be performed to characterise the ADC at different frequencies.

This test method does not allows for the characterisation of the converter at the transfer function level. Thus, e.g., errors associated with existing missing codes will be completely diluted within the remaining errors (noise, quantisation, jitter). The parameters extracted from spectral analysis are in fact affected by these transfer function localised errors but will not allow their identification neither to evaluate the ADC-UT behaviour at each single code.

2.2.2   Sine wave Fitting Analysis.   This method calculates a sinewave $\hat{x}[n]$ that best fits the sample record using a square error minimisation criterion . This calculation is performed by solving some nonlinear equations through an iterative algorithm. The fitting error (residue) is then calculated after the difference between the data record and the best fitting sinewave ($\hat{\eta}[n] = y[n] - \hat{x}[n]$) and used to characterise the performance of the ADC. The actual squared error is given by $\sigma_\eta^2 = \sum_{n=1}^{M} [y[n] - A\cos(\omega \times t_n + \phi) - C]^2$, where $y[n]$ represents the data sample record, and A, ω, ϕ, and C are the fitted parameters of, respectively, amplitude, frequency, phase, and offset. To find the best fitting sinewave the values of A, ω, ϕ, and C have to be found.

This approach provides a global behavioural description test since all the errors the test measures are averaged together. The standard deviation of these errors is then compared to the one which would be given by an ideal ADC of the same number of bits. A global evaluation of the ADC transfer function is provided through assessment of the effective number of bits (N_{ef}) figure of merit given by

$$N_{ef} = N - log_2 \frac{actual\ rms\ error}{ideal\ rms\ error} = N - log_2 \frac{\sigma_\eta}{\sigma_{\varepsilon_q}} \qquad (6.4)$$

The ideal *rms* error, assuming an ideal converter, with no linearity errors and an uniform probability density of the input stimulus, is given by $\sigma_{\varepsilon_q} = \frac{Q}{\sqrt{12}}$. When using a sinewave, the deviation from a uniform probability density over each single code bin is very small, except near the extremes, so that σ_{ε_q} does not appreciably deviate from the above value. This is particularly true for a high number of bits — for $N \geq 12$ the accuracy is better than 1% [9]. A plot of N_{ef} as a function of the input frequency can be obtained by calculating the effective bits for a series of frequencies spanning the useful frequency range of the ADC-UT.

From the variance of the residue it is also possible to obtain $SINAD = 10log\frac{A^2/2}{\sigma_\eta^2}$. More elaborated algorithms allow to get estimates also for SNR and THD [9]. In these algorithms the data record is approximated by the sum of a fundamental frequency sinewave with other sinewaves at frequencies multiple of the fundamental corresponding to the *h* first harmonics. The algorithm becomes however more time consuming and of more difficult convergence.

Besides requiring an input sinusoidal stimulus of high purity, these methods present some drawbacks due to the iterative nature of the algorithms, such as

- convergence is not guaranteed due, e.g., to very poor data or insufficient computational resolution — for the algorithm presented in [26] a precision $\leq$ 16-bit is often inadequate to guarantee convergence

- the results from different tests may not be consistent due to possible trapping at a local minimum [149]

- a long execution time is required to improve accuracy and convergence performance

- harmonic distortion can cause the parameter estimates to be biased because time truncated sinusoids of different frequencies are in general not strictly orthogonal

The efficiency of the method, in terms of accuracy, is limited by the following aspects

- a too large ratio between the sampling frequency and the stimulus frequency does not allow to memorize a sufficient number of samples

- in practice sometimes it is not possible to know in which quadrant of the sinewave the sampling point under process is located

- using an input frequency that is a submultiple of the sampling frequency violates the local uniform probability distribution assumption [26]

- the input stimulus amplitude must be carefully controlled otherwise clipping occurs and large errors may result. IEEE 1057 standard recommends using an amplitude ranging between 90% and 99% of V_{fs}; clipping together with adequate data processing is however recommended in [61] for better accuracy

Closed-form expressions linking the accuracy to the parameters describing measurement conditions have been derived [21]. It was shown that accuracy depends mainly on the ratio between the number of acquired samples and the number of ADC quantisation levels. Systematic errors are introduced by inaccurate synchronism between the generator and the digitiser and by incoherent sampling [118].

2.2.3   Comment on the calculation of noise power.   Generally, one say that the SA and SF methods evaluate the behaviour of ADC-UT by computing the noise captured at its output, i.e., both SA and SF methods provide an estimate of the ADC parameters from the estimated error sequence $\hat{\eta}[n] = y[n] - \hat{x}[n]$. This error results actually from both test setup noise and the ADC intrinsic noise. The variance of the overall conversion error, which corresponds to the total noise power, is given by $\hat{\sigma}_\eta^2 = 1/(M-1) \sum_{m=1}^{M} \hat{\varepsilon}_q^2[n]$. As the records include only a limited number of samples and input signal periods, random additive noise and timing jitter, cause the estimated parameters to be themselves random variables with associated variances. In [21] it is stated that, providing high statistical efficiency algorithms are used, the effect of the noise power of uncertainties in the estimated sinewave parameters can be neglected, and thus an unbiased estimate is obtained, even if M instead of $M-1$ is used in the previous equation.

Using SA and SF methods any variation among repeated measurements on the same signal should be due to random noise . The $(1-\epsilon) \times 100\%$ confidence interval within which the expected value of noise power is located is given by [59]

$$\frac{(M-1)\hat{\sigma}_\eta^2}{\chi_{\epsilon/2}^2} \leq \sigma_\eta^2 \leq \frac{(M-1)\hat{\sigma}_\eta^2}{\chi_{1-\epsilon/2}^2} \tag{6.5}$$

where χ^2 represents the chi-squared distribution for (M-1) degrees of freedom. The probability of having a measurement outside this range is ϵ. Similarly, the $(1-\epsilon) \times 100\%$ confidence interval for the SINAD measurement is [59]

$$\frac{\chi^2_{1-\epsilon/2}\sigma^2_s}{2\hat{\sigma}^2_\eta(M-1)} \leq SINAD \leq \frac{\chi^2_{\epsilon/2}\sigma^2_s}{2\hat{\sigma}^2_\eta(M-1)} \tag{6.6}$$

If R independent records of the ADC output response are taken, the sample mean ($\langle\hat{\sigma}^2_\eta\rangle$) of these R estimates of quantisation noise gives us the accuracy of the test method, and their variance provides its precision. Usually this precision is given by the variance of the R values $\hat{\sigma}^2_\eta(i)$ obtained, i.e.

$$\hat{\sigma}^2_{\langle\hat{\sigma}^2_\eta\rangle} = \frac{1}{R-1}\sum_{r=1}^{R}[\hat{\sigma}^2_\eta(i) - \langle\sigma^2_\eta\rangle]^2 \tag{6.7}$$

However, this process can be considered as a nested one, i.e., one wants to calculate the bias and variance of R measures, each one defined by an average value $\mu_\eta(i)$ and a standard variation $\sigma_\eta(i)$ and thus the precision of the measure should be given by the value S_η calculated according to the procedure illustrated in figure 6.2.

$$\begin{bmatrix} \eta_{11} & \eta_{12} & \cdots & \eta_{1j} & \cdots & \eta_{1M} \\ \eta_{21} & \eta_{22} & \cdots & \eta_{2j} & \cdots & \eta_{2M} \\ \vdots & \vdots & \ddots & \vdots & \ddots & \vdots \\ \vdots & \vdots & \ddots & \eta_{ij} & \ddots & \vdots \\ \vdots & \vdots & \ddots & \vdots & \ddots & \vdots \\ \eta_{R1} & \eta_{R2} & \cdots & \eta_{Rj} & \cdots & \eta_{RM} \end{bmatrix} \rightarrow \begin{vmatrix} \mu_\eta(1), \sigma^2_\eta(1) \\ \mu_\eta(2), \sigma^2_\eta(2) \\ \vdots \\ \mu_\eta(i), \sigma^2_\eta(i) \\ \vdots \\ \mu_\eta(R), \sigma^2_\eta(R) \end{vmatrix}$$

$$\downarrow$$

$$|[\mu_\eta, \sigma^2_\mu], [\langle\sigma^2_\eta\rangle, \sigma^2_{\langle\sigma^2_\eta\rangle}]|$$

$$\downarrow$$

$$|\mu_\eta, S^2_\eta = \sigma^2_\mu + \frac{\sum_{i=1}^{R}\sigma^2_\eta(i)}{R}|$$

η_{ij} – error of sample j in record i
$\mu_\eta(i)$ – average of M sample errors in record i
$\sigma^2_\eta(i)$ – variance of M sample errors in record i
μ_η – average of all errors over the R records
σ^2_μ – variance of the averages calculated for each record
$\langle\sigma^2_\eta\rangle$ – expected value from the variances of each record
$\sigma^2_{\langle\sigma^2_\eta\rangle}$ – variance of the variances calculated for each record
S^2_η – expected variance for the M sample errors over R records

Figure 6.2.   Arrangement of R measures each one comprising M quantisation errors.

2.3   Code Histogram Analysis

Within this method a histogram of the ADC-UT output codes, that is, the graph of the number of occurrences (probability of occurrence) of each output code [9], is computed . The ADC-UT is evaluated by comparing the probability of occurrence of each code with the expected one, which is determined by the probability density function characteristic of the input stimulus. Since the probability density function of a sinewave is not constant over a period the resulting histogram has to be corrected. To avoid large differences in code probability that occurs at the sinusoidal peaks, an amplitude is chosen to slightly overdrive the full-scale range. Coherent sampling is mandatory in order to avoid that some codes end up not being stimulated due to sampling always at the same relative phase instants.

From the statistical analysis of the resulting histogram it is possible, with a certain level of confidence, to obtain the location of the code transition levels and therefore the transfer characteristic of the converter. Once the code transition levels are known, we can calculate the integral and differential non linearities (INL and DNL). Notice that known the transfer characteristic it is possible, by simulation and applying spectral analysis, to get all the other parameters. The measure of INL and DNL is actually a measure of the difference between the computed transfer function and the estimated ideal one (best straight line) of the ADC. Thus the value of INL or DNL is directly dependent on the criterion used to find this ideal transfer function.

Providing coherent sampling is guaranteed, the uncertainty affecting code transition levels is determined as discussed in chapter 5, sections 5.1.3 and 5.3. In the uncertainty expressions, the standard deviation of the contributions arising from count uncertainty, additive noise and phase noise decrease with $\sqrt{M}^{-1}$.

Concerning the ratio of stimulus and clock frequencies, to obtain a contribution to the variance of the number of counts in the cumulated histogram smaller than 0.25, and taking into account that it is required that $\frac{|\Delta r|}{r} \leq \frac{1}{2JM}$, being $r = \frac{f_{in}}{f_s}$ [9], the relative frequency error of these two generators shall be such that $|\frac{\Delta f_{in}}{f_{in}} - \frac{\Delta f_s}{f_s}| \leq \frac{1}{2JM}$. This error takes into account the finite resolution of the synthesizers and their stability.

2.4   Number of samples

The accuracy of all the three methods is directly affected by the number of samples captured from the ADC response . SA and SF methods require that at least one sample per code is present in the data record. Due to the varying slope of the input sinewave and considering that sampling occurs at constant intervals, acquiring 2^N samples in the full scale range of the ADC, does not guarantees acquiring a sample per output code. Depending on how sampling is performed, for instance, a 1024 point data record may exercise only half the codes in an 8-bit device, and clearly only stimulates a small fraction of the codes in a 12 bit device. To obtain a sample per code from an ideal N-bit ADC in a single input stimulus period, the maximum sampling period must be such that within this time increment, the maximum amplitude change of the input stimulus is equivalent to Q, providing its peak-to-peak amplitude is equal to the full scale range. In this case the minimum number of samples should be given by $M = \pi \times (2^N - 1)$. For a N-bit real ADC with a worst case differential non linearity DNL_{max} (expressed in LSB), as some of the code bin widths decrease, a minimum record size given by [9]

$$M = \frac{\pi 2^N}{1 - |DNL_{max}|} \tag{6.8}$$

is required. Figure 6.3 shows the variation of M with DNL_{max} for three different number of bit ADCs. Note that as DNL_{max} approaches 1 LSB the record size increases significantly. This computation implies that the sampling

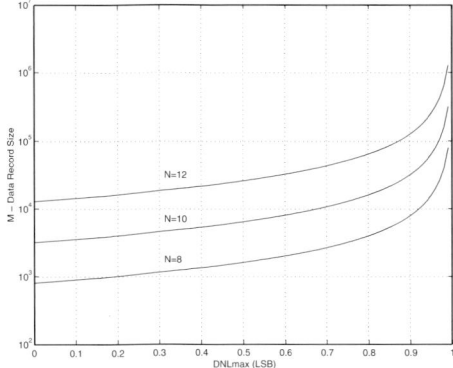

Figure 6.3. Variation of M with DNL_{max} for ADCs with 8, 10, and 12 bit.

instants are deterministic. If total jitter (stimulus phase noise, clock jitter, and ADC aperture uncertainty) is considered, the minimum record size required to capture at least one sample per code bin with a given probability p is determined by [9]

$$\frac{1}{2} erfc\left(\frac{\frac{\pi}{M_{min}} - \frac{1 - |DNL_{max}|}{2^N}}{2\pi f_i \sigma_J}\right) \geq p \tag{6.9}$$

where *erfc* represents the complementary error function and σ_J represents the worst case standard deviation of total jitter. Figure 6.4 shows, for a 12-bit converter with $DNL_{max} = 0.5$, the variation of the probability p with record size M for three jitter values.

When using SA for low frequency measurements, the required dynamic range also dictates the number of samples to be used. As the smallest achievable resolution bandwidth is as given in 6.3, for example an 8192 point FFT gives approximately a scale bin resolution of 6Hz at a 48kHz sampling rate.

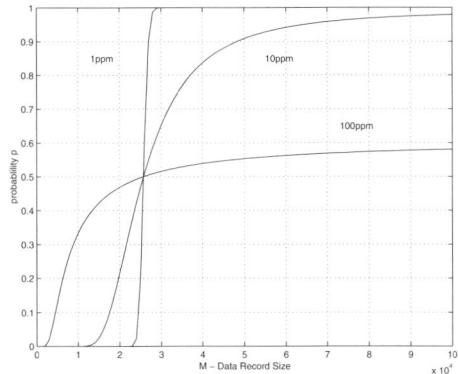

Figure 6.4.   Variation of probability p with record size M for three jitter values (1, 10 and 100 ppm).

To resolve a -100 dB 50 Hz spurious tone, requires a window attenuation of 100 dB only 5 bins away from the 20 Hz fundamental bin. This might not be feasible and require the data length to be increased.

In the histogram method the majority of the samples occur near the two ends of the histogram, therefore, it is not enough to have a single sample per code bin. In fact, to obtain each individual code transition level with a specified confidence ν that it does not deviate more then B LSB from the real value, we need a much higher number of samples per code bin, and therefore a longer data record. The minimum record size M, according to chapter 5, must be chosen such that

$$2 \left[\frac{2^{N-1} K_\nu}{B} \right]^2 \left[\frac{c\,\pi}{M} \right] \left\{ 1.13 \left[\frac{\sigma^*}{V_{rir}} + c\sigma_\phi \right] + \left[\frac{c\,\pi}{4M} \right] \right\} \le 1 \qquad (6.10)$$

for DNL, see (5.107), or that

$$\left[\frac{2^{N-1} K_\nu}{B} \right]^2 \left[\frac{c\,\pi}{M} \right] \left\{ 1.13 \left[\frac{\sigma^*}{V_{rir}} + \frac{c\sigma_\phi}{2} \right] + \left[\frac{c\,\pi}{4M} \right] \right\} \le 1 \qquad (6.11)$$

for INL, see (5.101), where the symbols are defined in chapter 5.

If no overdrive and noise are considered one obtains

$$B \le \frac{\sqrt{I}}{2} \frac{2^{N-1} K_\nu}{M} \frac{\pi}{\sqrt{R}} \qquad (6.12)$$

where I is 1 for INL and 2 for DNL.

Figure 6.5 shows the variation of M_{min} with uncertainty B (when the confidence level is $\nu = 0.9$) for three ADCs with different number of bits.

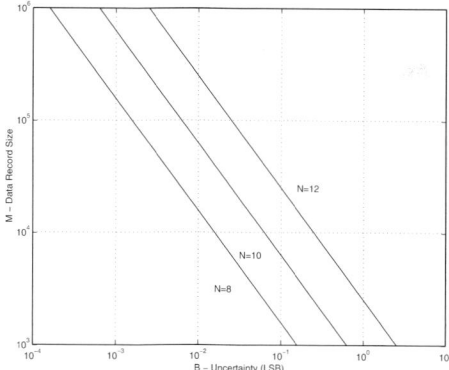

Figure 6.5. Variation of data record size M with uncertainty B for three different ADC resolutions (8, 10 and 12 bits).

Frequently the conditions represented by (6.10) or (6.11) and the condition

$$\frac{\Delta r}{r} \leq \frac{1}{2MJ} \qquad (6.13)$$

cannot be simultaneously met. In this case it is necessary to acquire an adequate number R of records with a maximum length M compatible with (6.13), as discussed in chapter 5.

2.5 Choice of test method

These three methods are usually considered to be ADC architecture independent methods, however the final application may dictate which method is the best for characterising ADCs to be used in that specific application . In fact, normally the generation of ADC tests is driven both by the application and by the ADC architecture, in order to choose the specification parameters to be measured, their limits, the test setup, or even the test methods themselves.

One of the critical aspects is the necessity of testing the converter as much as possible under the same functional conditions as those found in the final application. As this is not generally possible, a question is raised concerning the identification of the ADC parameters which best evaluate the ADC ability to perform the desired function. In general, the most important ADC performance characterisation parameter is SINAD. It is critical in almost all the applications.

In control applications the most critical ADC specifications concern the transfer function. In particular monotonicity, no-missing codes, and INL are the most important characteristics. Linearity for these ADCs is normally fully guaranteed. In fact, noise and harmonic distortion are not so important for these applications, provided linearity (both differential and integral) is guaranteed. In fact, while it is well known that THD and SNR are directly related to

INL and DNL, respectively, industrial processes correspond to a low-frequency range, where SNR changes very little with input frequency (and so does DNL) therefore SNR and THD are generally ignored in these applications.

For certain applications absolute conversion error may not be as important a characteristic as relative response variations to small variations of the input. This is the case, for instance, in digital audio applications, where nonlinearity results both in distortion of a single sinewave, described by THD, and in intermodulation effects, which downconvert high frequency tones to the baseband. The high resolution converters used in these applications require careful consideration of the clock jitter, for it causes a signal dependent modulation and additional noise. It is also important not to introduce group delay and gain differences which influence focus and stability of the sound sources in a stereo image. Performance at low levels and at higher frequencies is vital for good sound quality.

The key specification for wide-bandwidth spectral signal analysis is high linearity. SFDR is a fundamental limitation to the linear dynamic range in the spectral domain in order to ensure that the nonlinear behaviour of the ADC does not mask low-level narrowband signals. In radar applications the DR (dynamic range) is the key ADC performance, which ensures that the ADC noise does not affect the detection of targets. Both SFDR and DR are better measured using spectral analysis.

2.5.1   Measures provided by each method.   Table 6.1 summarizes the set of parameters obtainable with each method. Note that both SA and SF methods do not allow the characterisation of the converter at the transfer function level. Thus, errors associated with the existence of missing codes will be completely diluted in the overall behaviour of the ADC, and would then be seen as noise, quantisation error, or jitter. The parameters extracted by SA and SF methods are affected by these transfer function localized errors but will not allow their identification. Actually, methods have been recently proposed to obtain an evaluation of non-linearity after spectral analysis [15, 101, 128].

It is seen that SA provides a larger capability in the analysis of noise and distortion components than SF. In general, the values provided by SF using the 3-parameter method are in perfect agreement with those given by SA [21] when using coherent sampling. In the case of incoherent sampling the SF 4-parameter method presents a better performance.

With the SA method information on noise floor, harmonics, and spurious can be extracted. If all harmonics and spurious components are at least $6N$ dB below the full-scale amplitude of the fundamental, then the ADC-UT is performing satisfactorily since each error component has a peak-to-peak amplitude smaller than 1 LSB [26]. If, on the other hand, harmonic or spurious components are higher than $6N$ dB below full scale, or the noise floor is elevated (e.g.

Table 6.1. Parameters obtained with the three methods.

	SA	SF	CH
SINAD	$\checkmark$	$\checkmark$	‡
N_{ef}	$\checkmark$	$\checkmark$	‡
THD	$\checkmark$	†	‡
SNR	$\checkmark$	†	‡
SFDR	$\checkmark$		
INL			$\checkmark$
DNL			$\checkmark$

† — using more elaborated algorithms
‡ – after simulation of the obtained transfer function

due to code discontinuities), then other tests can be performed to identify INL or DNL errors. For instance, SA could be followed by a CH test.

Several applications require testing intermodulation distortion. As this test implies measuring the amplitude of beat frequencies generated by a two-tone input stimulus, SA is the most convenient method.

A CH test provides both a localized error description as well as some global description of the ADC. Detection of missing codes and measure of gain and phase at the test frequency can also be obtained with this method. It is thus a convenient method to test ADCs for control applications. However, even this may be fallacious. Codes that occur on positive going edges but not on the negative ones are not seen. Also, the sign of the quantisation error may change with the signal direction (a hysteresis effect) and this is not seen if a global INL or DNL curve is obtained [106]. As the CH method is based on the evaluation of the probability density function of the ADC output codes, it is insensitive to out-of-phase higher harmonic components at the ADC response, fact which can lead to nonlinearity underestimates.

The SF method is able to detect nonlinear defects (harmonic distortion), random noise, and aperture uncertainty. Gain, offset and phase errors do not affect the results since all the ADC errors related to parameters which are varied during the best fit procedure are not detected. While the apparent frequency error (due to time base frequency offset or drift of the ADC) is usually small and the DC offset can be assessed through a static calibration, amplitude distortion and phase nonlinearity are difficult to measure and are not easily removed [91].

The shortcoming of using the N_{ef} as a single measure of dynamic performance is, that it does not separate error components uncorrelated with the signal (noise) from correlated error (distortion). Although the end result of this process is a single figure of merit, some better understanding can be obtained by varying the test conditions [26]

- the randomness exhibited in the residues allows to identify noise

- white noise will produce the same degradation regardless of input frequency or amplitude — the error term will be independent of test conditions for this sort of error

- aperture uncertainty is identifiable because it generates an error that is a function of input slew rate. When this is the dominant error causing a low N_{ef}, the number of effective bits will scale with both input frequency and amplitude linearly. If the input waveform is sampled only at points of constant slew rate, such as at the zero crossings, then the aperture uncertainty may be correlated with the reduction in the effective bits as a function of slew rate.

- the amplitudes of the harmonics can be extracted by fitting the error residue with best fit sinewaves of the important harmonic frequencies. The impact of noise and aperture uncertainty in the presence of large distortion errors can be assessed by effective bit values and error residues after removing the fitted harmonics.

N_{ef} is sensitive to both the location and the magnitude of the nonlinearity errors, but it can not discern the two components. Solutions to minimize this drawback are discussed in [61]

- the amplitude and offset of the test sinewave are chosen in such a way as to stimulate the full-scale range of the ADC with some overdrive, while not saturating the analogue front-end of the ADC during the acquisition (peak to peak signal not higher than 120% of the full-scale)

- take a linearized (weighted) N_{ef} parameter defined in such a way as to be sensitive only to nonlinearity errors magnitude, but not to their location on the transfer characteristic

The implementation of the three methods might require specific approaches depending on the ADC-UT architecture. For instance, in the case of $\Sigma\Delta$ (oversampling) ADCs the noise floor is not as uniform as in other converters, thus an accurate characterisation of the noise floor might require spectral averaging using very long data records. Furthermore, the operation of these ADCs is intrinsically dynamic. In fact, the $\Sigma\Delta$ modulator can be thought of as a PCM converter with feedback, which attempts to force the output signal to be equal to the input stimulus. For this reason it tends to be inherently linear. Due to their high resolution, these converters require capturing a high number of samples, making the use of the CH method prohibitive.

2.5.2   Coherent sampling requirements.   All the three methods, SA, SF, and CH, require that not the same codes are sampled at exactly the same voltage in each cycle, otherwise the locally uniform probability distribution assumption is violated . To prevent this to occur the stimulus and sampling frequencies shall be related such that $M f_i = J f_s$, being J and M mutually prime integers. Additional requirements are discussed in the relevant chapters of this book: chapter 3 for SF, chapter 4 for SA, and chapter 5 for CH.

The use of windowing to overcome spectral leakage when perfect coherent sampling is not obtained is discussed in chapter 4, being the application of some common windows described in section 4.6.9.3.

Recently, the use of zero-order discrete prolate spheroidal sequences (DPSS) has been proposed as an alternative windowing procedure to obtain maximum measurement accuracy when non-coherency aplies [51]. These windows, which were firstly proposed by D. Slepian in 1978 [134], are finite length sequences that have the property of maximally concentrating the energy in the main lobe of the frequency response. They comprise the most spectral efficient set of orthogonal sequences possible and have maximum energy concentration in the frequency pass-band and minimum ringing in the time domain. The DPSS are thus optimal regarding their energy concentration in a given frequency sub-band. DPSSs are parameterized by the time bandwidth product MW, where W is the normalized one-sided bandwidth in Hz, and M is the number of samples. Choosing the time-bandwidth product to be $MW = 1.84$ gives a main lobe width in the frequency domain equal to that of the Hamming window. The Kaiser window [80] is an approximation to the zeroth order DPSS. The zeroth order window in the Slepian sequence is an excellent data tapering window. The subsequent windows in a Slepian series will emphasize the data better near the edges, but they do not offer the same spectral leakage resistance and they do not compute the true frequency. Instead, they contribute to the overall envelope of a multitaper spectral peak. There are two important factors corresponding to a window: its bandwidth, which determines how wide a frequency range it allows leakage over, and its rejection, which defines how much energy is re-duced when it is outside that frequency range. If the bandwidth W is small, the frequency range of the allowed leakage will be small. But a smaller W leads to a rejection degradation as more energy leaks outside the allowed range.

The class of windows proposed in [51] is defined in the frequency domain on the basis of the Dirichelet kernel as

$$W[k, \Lambda] \triangleq \frac{sin[\frac{M}{2} \arccos(\gamma \cos(2\pi k) + (\gamma - 1))]}{sin[\frac{1}{2} \arccos(\gamma \cos(2\pi k) + (\gamma - 1))]}, \quad k = 0, ..., M-1 \quad (6.14)$$

where $W[., \Lambda]$ represents the DFT of the M-sample window,

$$\gamma \triangleq (1 + \cos(2\pi/M)/(1 + \cos(2\Lambda\pi/M))) \quad (6.15)$$

and Λ is the mainlob width expressed in bins.

In case of non-coherent sampling an optimum Λ is obtained from

$$\Lambda_{opt} = 0.607 + 0.189 log_{10} M_\eta + 0.378 log_{10} \gamma_1 \qquad (6.16)$$

being $M_\eta \geq \frac{ENBW}{\epsilon^2}$ the number of samples associated with wideband noise — ϵ is the required estimator accuracy, but usually one can use $M_\eta \approx M/3$. An optimum number of samples associated to each narrow-band component, obtained after a compromise among maximum estimator accuracy, maximum frequency selectivity, and low computational effort, becomes $M_{X_i} = 2 \lceil \Lambda opt \rceil + 1$, where the operator $\lceil . \rceil$ rounds to the nearest integer, and making sure that

$$M > \frac{2 \lceil \Lambda opt \rceil + 1}{min_{i \neq j} |f_i - f_j|} f_s, \quad i, j = 1, h, sp \qquad (6.17)$$

i.e., the distance between the two closest narrow-band components is greater than M_{X_i} bins, for the fundamental, harmonics, and spurious components.

The window coefficients $w[n]$ are then calculated after substituting Λ_{opt} in (6.17) and by applying the inverse FFT to the resulting expression. It should be verified that $M_\eta \geq M - [2(nr_{harmonics} + nr_{spurious}) + 1] M_{X_i}$ and that the required accuracy is attained, otherwise the number of samples has to be increased. The reader is exorted to see the published bibliography [30, 51, 116]

3.   Simulation results

3.1   Simulation procedure

Simulations were carried-out in order to compare the three methods presented in previous chapters, and to evaluate how they behave in the presence of non-ideal conditions. Results were obtained to evaluate the minimum number of samples required by each method, the influence of total noise, harmonic distortion, and non-coherent sampling in the accuracy and precision of the methods, as well as how these are affected by the presence of offset. Finally, the methods were compared in terms of testing time.

The following procedure was adopted. Simulations were performed with MATLAB using the polynomial ADC transfer characteristic models depicted in Table 6.3.1 [45].

Each parameter estimation was obtained from the results of 30 simulations, using the average of these 30 values to calculate the expected value (accuracy), and their standard deviation to evaluate the precision of the estimation. The variability of the test results is obtained by introducing randomness in the initial phase of the stimulus sinewave. The resulting data records where then processed using each one of the three test methods to calculate the SINAD parameter. The choice of SINAD as a figure of merit is due to the fact that it is possible to calculate it with the three methods. All results were obtained for ideal (ADC$_1$)

Table 6.2. ADC model used in the simulations.

$h(x) = x_0 + \alpha_1 x + \alpha_2 x^2 + \alpha_3 x^3 + \alpha_4 x^4 + \alpha_5 x^5$	
ADC$_1$	***ADC$_2$***
$\alpha_0 = 0$	$\alpha_0 = 0$
$\alpha_1 = 1$	$\alpha_0 = 1$
	$\alpha_2 = 0.5 LSB$
	$\alpha_3 = 1 LSB$
	$\alpha_4 = 0.25 LSB$
	$\alpha_5 = 0.75 LSB$

and non-ideal (ADC$_2$) models, each one parameterized for 8 and 12 bits. Figure

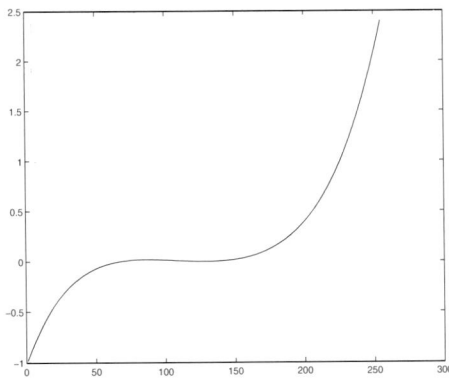

Figure 6.6. INL curve of the non-ideal converter (ADC$_2$).

6.6 shows the INL curve which results for the non-ideal model ADC$_2$.

3.2 Results using SA and SF

3.2.1 Number of samples. In order to find a minimum number of samples which guarantees that the accuracy of the estimates is not affected by the number of samples being used, SINAD values were obtained using both SA and SF for different values of M. Table 6.3 shows the mean and standard deviation values obtained for the four ADC models. It can be seen that for $M \geq 16384$ the precision in the estimation of SINAD is, in any case, smaller than .05 dB. This guarantees that results can be taken accurate up to the first decimal number. Using this indicator all results were afterwards obtained using 16384 samples.

The figures in Table 6.7 illustrate the results obtained for the ideal ADC cases. It can be seen that for the 8-bit cases 1024 samples ($\log_2(M) = 10$)

Table 6.3.   SINAD estimation accuracy and precision for different number of samples.

SA

	8 bits				12 bits			
	ADC_1		ADC_2		ADC_1		ADC_2	
M	mean	std	mean	std	mean	std	mean	std
256	49.2594	0.3844	47.4018	0.2924	73.9272	0.3176	71.8784	0.3706
512	49.2002	0.0663	47.4685	0.1874	73.7254	0.2554	71.8412	0.2602
1024	49.1810	0.0218	47.4089	0.0277	73.8537	0.1353	71.8997	0.1947
2048	49.1918	0.0060	47.4044	0.0212	73.7755	0.1636	71.8548	0.1449
4096	49.1932	0.0031	47.4096	0.0100	73.7767	0.1099	71.8722	0.0977
8192	49.1957	0.0023	47.4124	0.0095	73.7890	0.0559	71.8861	0.0349
16384	49.1973	0.0017	47.4126	0.0028	73.7977	0.0071	71.8823	0.0265
32768	49.1971	0.0007	47.4132	0.0015	73.7966	0.0062	71.8854	0.0173
65536	49.1974	0.0003	47.4134	0.0007	73.7971	0.0005	71.8854	0.0031

SF

	8 bits				12 bits			
	ADC_1		ADC_2		ADC_1		ADC_2	
M	mean	std	mean	std	mean	std	mean	std
256	49.3294	0.3830	47.4767	0.2954	73.9991	0.3162	71.9513	0.3667
512	49.2357	0.0675	47.4882	0.1821	73.7617	0.2555	71.8704	0.2622
1024	49.1988	0.0222	47.4197	0.0271	73.8720	0.1355	71.9093	0.1947
2048	49.2010	0.0059	47.4109	0.0222	73.7844	0.1636	71.8596	0.1444
4096	49.1980	0.0039	47.4134	0.0101	73.7813	0.1098	71.8746	0.0977
8192	49.1983	0.0023	47.4146	0.0099	73.7912	0.0559	71.8874	0.0349
16384	49.1993	0.0018	47.4142	0.0030	73.7988	0.0071	71.8830	0.0265
32768	49.1977	0.0006	47.4136	0.0015	73.7972	0.0062	71.8858	0.0174
65536	49.1977	0.0003	47.4136	0.0007	73.7974	0.0005	71.8856	0.0031

would be sufficient to guarantee the required accuracy. This corresponds to an $DNL_{max} = 0.22$ as given by (6.8). The values obtained with the two methods are similar.

3.2.2 Effects of noise, harmonic distortion and non-coherence.
Another set of simulations was performed to evaluate the effect of additive noise . Note that besides modelling real additive noise superimposed to the input signal, additive noise provides a rough model of the effect of the h.f. phase noise of the generators and of intrinsic sampling jitter.

The presence of noise was modelled by "injecting" into the test stimulus a pseudo-random gaussian noise generated with the MATLAB command *randn*, with an amplitude power in dB as given in the first column of Table 6.4. This table presents the values obtained under these conditions. The ideal values are presented in the last line of the table.

Again, it can be seen that whether using SA or SF, the results are similar and so one can conclude that they behave similarly in the presence of noise. Also, as expected, for noise levels (absolute value) higher than the SINAD of the ideal noiseless ADC, the measured SINAD values are significantly affected, i.e., while a 8-bit ADC tolerates a -60 dBc noise level without significant degradation of its SINAD, a 12-bit ADC requires that the noise level does not exceed -80 dBc. This is in accordance with the $\simeq -6N$ limit given for an ideal ADC. For additive noise levels below these, the ADC intrinsic quantization noise dominates, and obviously the SINAD of the non-ideal ADC is worse than the one of the ideal ADC. On the other hand, for higher levels of the additive noise, the stimulus noise dominates and one can not distinguish the ideal from the non-ideal ADC. This is true for both methods.

The effect of harmonic distortion in the input signal is demonstrated by the results in table 6.5. These values were obtained considering an input sinewave affected only by 3rd harmonic distortion. Besides the similarity of results given by the two methods, it can be seen that distortion amplitudes smaller than the ideal SINAD values do not affect the estimation accuracy. One can notice that the standard deviation presented by the non-ideal ADC (ADC_2) is much higher than that presented by the ideal one (ADC_1), particularly for the 8-bit case. This is due to the fact that, in this specific case, 3rd order harmonic distortion is also introduced by the ADC itself. The maximum standard deviation (ADC_2) occurs in the region where stimulus harmonic distortion is of the same order of magnitude (absolute values) of the ADCs' ideal SINAD — -50 dBc for the 8-bit ADC, and -74dBc for the 12-bit one. For values below these, the harmonic distortion introduced by the ADC dominates, the SINAD values obtained for the non-ideal ADC being smaller than the ones of the ideal ADC. On the other hand, for values above those the harmonic distortion of the stimulus dominates and one can not distinguish the ideal from the non-ideal ADC. This is valid for

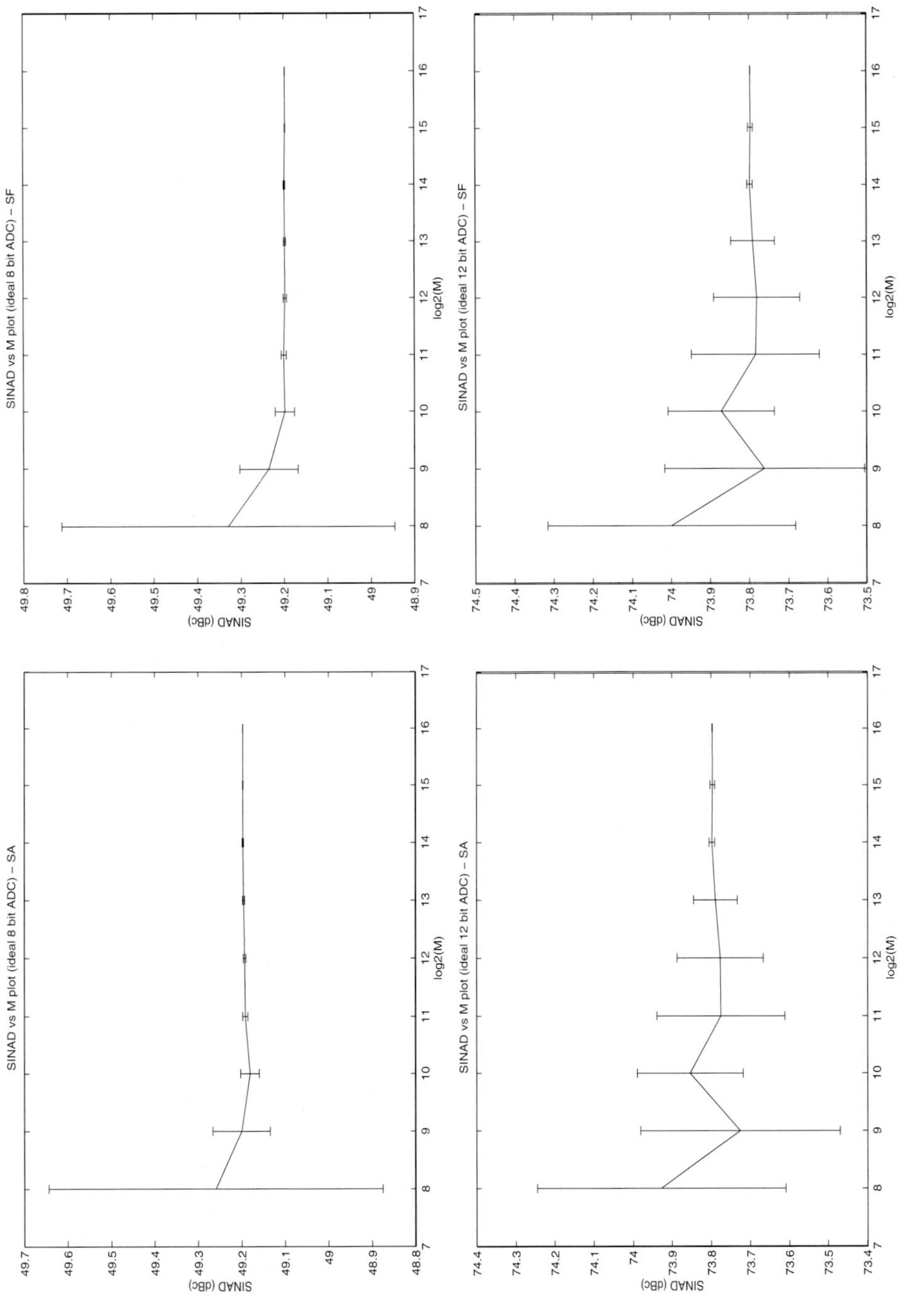

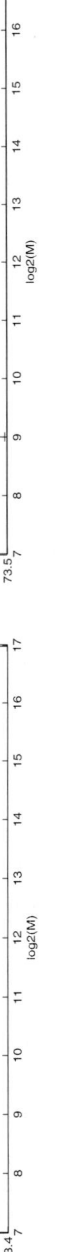

Figure 6.7.   SINAD estimation in 8- and 12-bit ideal ADCs for different number of samples.

Table 6.4. SINAD accuracy and precision as a function of the noise level.

SA

Noise (dBc)	8 bits				12 bits			
	ADC_1		ADC_2		ADC_1		ADC_2	
	mean	std	mean	std	mean	std	mean	std
-40	36.9950	0.0470	36.8455	0.0583	37.1838	0.0541	37.1931	0.0428
-60	48.5618	0.0189	46.9827	0.0207	56.9607	0.0572	56.8944	0.0477
-80	49.1900	0.0029	47.4073	0.0069	72.1119	0.0363	70.7392	0.0398
-100	49.1969	0.0021	47.4126	0.0026	73.7761	0.0081	71.8712	0.0229
$-\infty$	49.1973	0.0017	47.4126	0.0028	73.7977	0.0071	71.8823	0.0265

SF

Noise (dBc)	8 bits				12 bits			
	ADC_1		ADC_2		ADC_1		ADC_2	
	mean	std	mean	std	mean	std	mean	std
-40	36.9967	0.0471	36.8473	0.0584	37.1856	0.0539	37.1951	0.0428
-60	48.5632	0.0191	46.9837	0.0208	56.9616	0.0571	56.8955	0.0478
-80	49.1919	0.0030	47.4086	0.0068	72.1130	0.0362	70.7399	0.0398
-100	49.1985	0.0023	47.4141	0.0030	73.7771	0.0081	71.8719	0.0229
$-\infty$	49.1993	0.0018	47.4142	0.0030	73.7988	0.0071	71.8830	0.0265

both methods and allows us to conclude that the cost of filters to be used in the test setup would be similar.

Finally, table 6.6 describes the effects of non-coherent sampling. The level of non-coherence is given by the number of cycles inaccuracy ϵ_J : $f_i = (J \pm \epsilon_J)\frac{f_s}{M}$. In this table *mod()* represents the *modulus* (signed remainder after division) operation. A Blackman-Harris (7 terms) window is used in the first four lines. The fifth line corresponds to the ideal case, i. e., coherent sampling and no windowing used. It can be seen that, provided windowing is used, no significant degradation can be observed in the results obtained by the SA method in comparison with those obtained by SF (which is not affected by repeated sampling at the same relative phase instants, once an integer number of cycles is captured).

Concerning non-coherent sampling, there is no definite advantage in using one method or the other, although it should be remarked that SA requires windowing and SF presents always smaller standard deviations, even when non-coherence is small. SF is, thus, not significantly affected by non-coherence.

3.2.3   Testing time.   Testing time was evaluated in terms of both data acquisition time and data processing time . For SF and SA data acquisition time is about the same as the number of acquired samples is the same. Table 6.7 presents data processing time values obtained for these two methods. Time is computed using the command *cputime* and the number of flops using *flops*.

Data processing time is directly determined by the number of samples, but not by the number of bits of the ADC. In fact, with the current 32-bit PCI data bus architectures, the memory access and arithmetic operations are the same for any number of bits of the current ADCs. Data processing time is higher for the SF method because the number of floating point operations is higher. Within this method the iterative process is stopped when the difference between parameter matrix values at iteration $n+1$ do not differ more than 1×10^{-6} from the previous ones. This was considered a good criterion as the precision values obtainedwith SF are similar to those obtained with SA. The algorithms used require a number of floating point operations given by $flop = K_f M log_2 M$ in both cases, but the constant of proportionality k_f is significantly higher for the SF method. In fact, it should be borne in mind that in the SF method the number of flops depends on the precision value which determines the end of the iterative process. Anyway, it is worth to mention that in general 3 iterations are enough to stop the process, and after 2 iterations the difference between matrix values is about 1×10^{-3}. As far as processing time is concerned, it cannot be proportionally related with the factor $M log_2 M$.

3.2.4   Effects of stimulus amplitude and offset.   It was mentioned in section 6.2.2.1 that the presence of offset in the stimulus may affect THD and

Table 6.5. SINAD accuracy and precision as a function of the stimulus 3rd harmonic content.

SA

	8 bits				12 bits			
	ADC_1		ADC_2		ADC_1		ADC_2	
3rd harmonic (dBc)	mean	std	mean	std	mean	std	mean	std
-40	39.6710	0.1944	39.6860	1.2079	40.1735	0.2580	40.2081	0.3351
-60	48.8693	0.0287	47.1444	0.5040	59.8741	0.0662	59.8271	0.7930
-80	49.1929	0.0033	47.4006	0.0476	72.8755	0.0218	71.4010	0.9168
-100	49.1972	0.0020	47.4119	0.0064	73.7849	0.0068	71.8883	0.1139
$-\infty$	49.1973	0.0017	47.4126	0.0028	73.7977	0.0071	71.8823	0.0265

SF

	8 bits				12 bits			
	ADC_1		ADC_2		ADC_1		ADC_2	
3rd harmonic (dBc)	mean	std	mean	std	mean	std	mean	std
-40	39.6717	0.1944	39.6867	1.2078	40.1740	0.2580	40.2086	0.3350
-60	48.8709	0.0289	47.1457	0.5037	59.8742	0.0662	59.8271	0.7931
-80	49.1948	0.0031	47.4021	0.0475	72.8764	0.0218	71.4017	0.9170
-100	49.1992	0.0021	47.4133	0.0063	73.7860	0.0068	71.8893	0.1139
$-\infty$	49.1993	0.0018	47.4142	0.0030	73.7988	0.0071	71.8830	0.0265

Table 6.6.   SINAD accuracy and precision as a function of non-coherence level.

SA

	8 bits				12 bits			
	ADC_1		ADC_2		ADC_1		ADC_2	
$mod(f_i/(f_S/M))$	mean	std	mean	std	mean	std	mean	std
0.1	49.1916	0.0048	47.4077	0.0255	73.7940	0.0421	71.9054	0.0793
0.01	49.2016	0.0156	47.4266	0.0256	73.7958	0.0468	71.8753	0.0677
0.001	49.1710	0.1654	47.4236	0.0284	73.8109	0.0656	71.8935	0.0598
0	49.1932	0.1167	47.4227	0.0193	73.7952	0.0323	71.8761	0.0427
0, no wind.	49.1973	0.0017	47.4126	0.0028	73.7977	0.0071	71.8823	0.0265

SF

	8 bits				12 bits			
	ADC_1		ADC_2		ADC_1		ADC_2	
$mod(f_i/(f_S/M))$	mean	std	mean	std	mean	std	mean	std
0.1	49.1983	0.0062	47.4140	0.0069	73.7951	0.0098	71.8950	0.0316
0.01	49.1993	0.0128	47.4207	0.0191	73.7971	0.0162	71.8733	0.0389
0.001	49.2064	0.0420	47.4134	0.0084	73.7993	0.0154	71.8975	0.0235
0	49.1993	0.0018	47.4142	0.0030	73.7988	0.0071	71.8830	0.0265

Table 6.7. Testing time.

SA	8 bits		12 bits	
M	**time (s)**	**flop**	**time (s)**	**flop**
256	0.09	10420	0.08	10420
512	0.11	21615	0.11	21615
1024	0.16	45239	0.16	45239
2048	0.25	94968	0.26	94968
4096	0.47	199498	0.46	199498
8192	0.86	418733	0.86	418733
16384	1.69	877593	1.67	877593
32768	4.03	1836174	3.34	1836174
65536	7.09	3835148	6.79	3835148
SF	**8 bits**		**12 bits**	
M	**time (s)**	**flop**	**time (s)**	**flop**
256	0.26	62679	0.26	62679
512	0.45	125076	0.45	125076
1024	0.85	251100	0.85	251100
2048	1.63	505645	1.62	505645
4096	3.19	1019783	3.20	1019783
8192	6.39	2058218	6.40	2058218
16384	12.89	4155478	12.87	4155478
32768	31.07	8390859	25.86	8390859
65536	52.81	16943433	51.79	16943433

SFDR measurements accuracy due to the effects on the quantisation of the sinewave peaks. A different set of simulations was performed to evaluate this effect. Graphics in figure 6.8 present the evolution of THD, respectively SINAD for 6- and 12-bit ADCs, as a function of the stimulus relative amplitude. Dotted lines represent the results obtained with the typical setup, while continuous lines represent the case where "wobbling" is added. Wobbling consists of adding to the input stimulus an offset value which increases along the stimulus duration from *-1/2 LSB* to *1/2 LSB* [45]. It can be seen that the accuracy of the THD and SINAD check results can be significantly affected using the typical setup. By adding "wobbling", a better distribution of the input stimulus along the ADC-UT input scale is obtained, leading thus to a better accuracy and precision of the results. Notice that a variation of 2% in the stimulus amplitude can lead to a variation of about 0.8dB in the SINAD value. It is also curious to note that the maximum SINAD occurs for a stimulus amplitude smaller than the full scale amplitude, and different from the amplitude that gives the best THD (top row graphics).

Of course this effect is more pronounced in low resolution ADCs, such as those used in video applications. The second row of figures in Table 6.8 shows the results obtained with a 12-bit ADC. Although some differences are still observable in THD, that is not the case in the SINAD plot.

3.3   Results using CH

The CH method was evaluated by estimating the values of THD and SINAD for different values of uncertainty in the estimation of the code transition levels (parameter B in (5.99)). Graphics in figure 6.9 show the evolution of these two parameters. As it would be expected, the uncertainty on the measurements increases with the uncertainty accepted on the estimation of code transition levels. Using again the criterion of accepting as maximum uncertainty the one that allows an accuracy on the parameters estimation up to the first decimal number, an error of 0.025 LSB would be acceptable on the estimation of the code transition levels.

Table 6.8 summarizes the uncertainty values obtained on the definition of the code transition levels. The three columns of the total number of samples (S) result from 10, 100, and 1000 records, and 1024 samples per record. In each case, the uncertainty was found after the standard deviations of the transition levels obtained by simulation comparing to the ideal values given by the curve shown in figure 6.6. It can be seen that the observable decrease in uncertainty is proportional to the square-root of the increase in the number of samples, as stated before.

Comparing the SINAD values presented in figure 6.9 obtained by using the CH method, with those presented in figure 6.7, both for the 8-bit ADC_2 case,

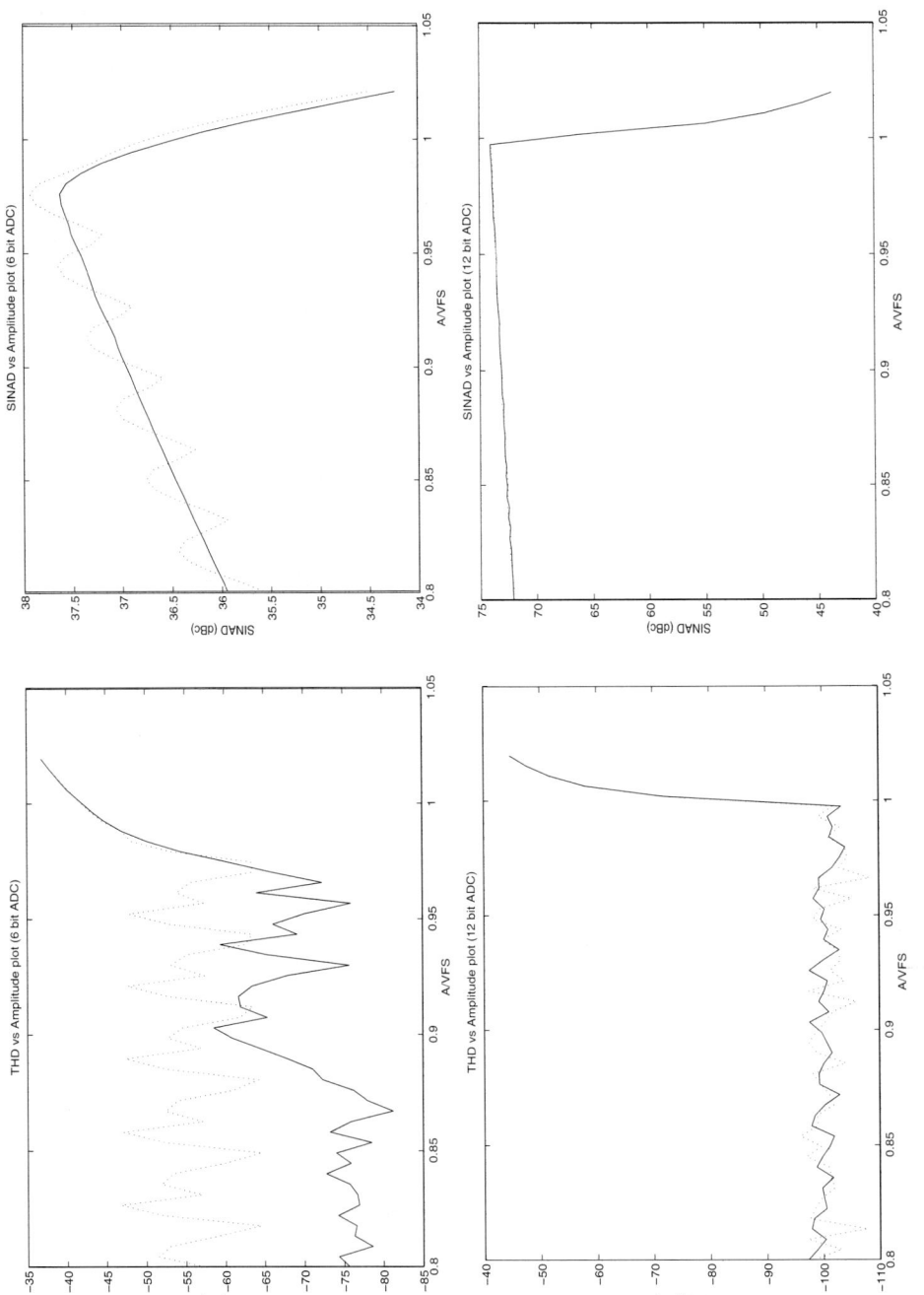

Figure 6.8.   Variation of THD and SINAD with the stimulus amplitude.

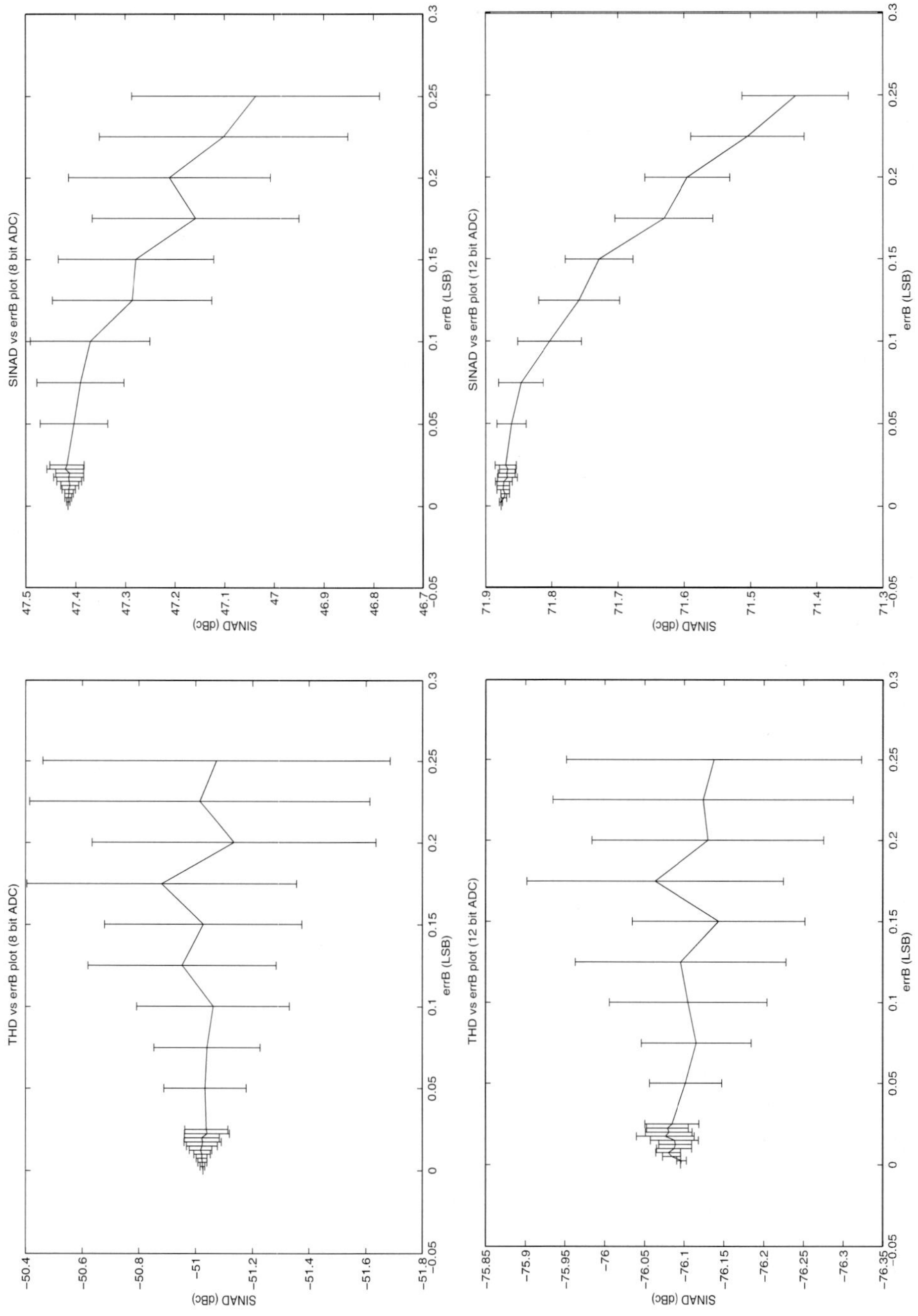

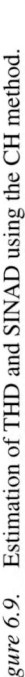

Figure 6.9.   Estimation of THD and SINAD using the CH method.

Table 6.8. Code transition levels estimation uncertainty.

N	8	8	8	12	12	12
S=R×M	10240	102400	1024000	10240	102400	1024000
$B_{simulation}$	0.0636	0.0232	0.0075	1.1927	0.4158	0.1172

one can see that the precision obtained with an uncertainty $B = 0.025 \rightarrow S = 102400$ is similar to the one obtained with $log_2(M) = 10 \rightarrow M = 1024$ samples using the SA or SF methods. In the 12-bit case, to obtain the same precision obtained with $log_2(M) = 14 \rightarrow M = 16384$ samples within the SA and SF methods, the CH method would require more than 1024000 samples. A larger number of samples is thus required by the CH method than the SA or SF ones to obtain the same measurement accuracy. It can also be seen that outside these limits all the methods (CH, SA and SF) demonstrate a significant degradation both in terms of accuracy and precision.

3.3.1 Effects of noise, harmonic distortion, and non-coherence.

Table 6.9 shows the effects of additive noise, harmonic distortion, and non-coherence on the performance of the CH method . $S = R \times M = 1000 \times 1024 = 1024000$ samples are used in these evaluations. Only results for ADC_1 are presented. The first fact one can remark is that the degradation of the SINAD estimation is smaller using the CH method. In the 8-bit case practically no effects are observable, while in the 12-bit one only in the worst case perturbations are noticeable. This is true for all the three perturbations studied, and indicates that CH is more robust than the other methods, particularly concerning the non-coherent sampling situation where the difference between maximum and minimum values is 0.0059 dB and 1.2853 dB, respectively, the 8- and 12-bit cases.

3.3.2 Testing time.

Concerning testing time, although the CH method requires a smaller data processing time, the higher number of samples makes its data acquisition time significantly higher than the one required by the other two methods . This can be seen in table 6.10 which was obtained for a 10-bit ADC with a sampling frequency of 100 kHz.

4. ATE Implementation

This section provides examples of implementation of the three test methods on a mixed-signal ATE, together with some experimental results .

Table 6.9. SINAD accuracy and precision as a function of the noise level, harmonic distortion and non-coherence.

CH	*ADC$_1$*			
	8 bits		*12 bits*	
Noise (dBc)	*mean*	*std*	*mean*	*std*
-40	49.6784	0.0195	66.3306	0.1354
-60	49.9580	0.0005	73.4600	0.0650
-80	49.9587	0.0011	73.7324	0.2173
-100	49.9587	0.0013	73.7594	0.1656
-∞	49.9588	0.0008	73.8378	0.1343

CH	*ADC$_1$*			
	8 bits		*12 bits*	
3rd harmonic (dBc)	*mean*	*std*	*mean*	*std*
-40	49.7118	0.1553	67.8120	1.7236
-60	49.9584	0.0007	72.9426	0.3336
-80	49.9587	0.0008	73.7969	0.1951
-100	49.9591	0.0006	73.7501	0.2073
-∞	49.9588	0.0008	73.8378	0.1343

CH	*ADC$_1$*			
	8 bits		*12 bits*	
$mod(f_i/(f_s/M))$	*mean*	*std*	*mean*	*std*
0.1	49.9529	0.0043	72.5525	0.8171
0.01	49.9539	0.0063	72.5161	1.1750
0.001	49.9579	0.0020	73.6063	0.2011
0	49.9588	0.0008	73.8378	0.1343

Table 6.10. Data acquisition and processing times required by the three methods to test a 10-bit ADC.

Time (s)	*Acquisition*	*Processing*	*Total*
CH	6.55	0.6	6.61
SA	0.08	0.99	1.07
SF	0.08	5.34	5.42

4.1 Features

Use of ATE systems is recommended in production environment where a large number of devices of the same model has to be tested in order to verify the compliance of their electrical characteristics to the specifications provided. The main feature of such macro-systems is the integration of a lot of measurement sub-systems, making them available, through appropriate cables, to the test board of the device in the form of "pogo pins" (i.e. measurement points very close to the device to be tested). The noise and the interference of the sub-systems can be controlled and measured by the macro-system, which, in turn, can run proper calibration programmes in order to reach a very good performance in terms of accuracy of the measurement and of noise on the test board. The memory dedicated to data acquisition is usually local (per pin memory) and thus does not use the same memory dedicated to the calculations. Another advantage is the software control of all the measurement sub-systems and their synchronization. The drawbacks are the high costs of both the hardware — as the test boards have to be designed in compliance with the test-head specifications — and software, that is written in a dedicated language and thus requires proper personnel training.

4.2 Software development

As example of implementation of the algorithms proposed, a test programme was developed on a Synchromaster AC - LTX ATE system. The DUT (Device Under Test) is an 8-bit A/D converter with a typical sampling frequency of 50 MHz. The operating conditions for the DUT were chosen to be: $M = 16384$, $f_s = 50053120.0$ Hz, $f_{in} = 21088665.0$ Hz, ($J = 6903$ - in order to satisfy the coherence condition.)

A programme developed on an ATE allows one to test all electrical parameters and compare them to the specification limits giving a PASS or FAIL, the later result used to discard bad devices. Moreover a data-log file is usually associated to each run of the programme: this maintains a hardcopy of the measurement results for the parameters obtained from each tested device. The main target for a good test engineer is to cover as accurately as possible, and as exhaustively as possible, the electrical performance of the device to be tested and to do it as fast as possible: in other words, test time and test coverage are the main issues to be addressed. Herein follows an example of the SW implementation to test an ADC according to the three methods exposed in previous chapters. It is given in a simple pseudo-code format. It should be kept in mind that in reality tricks are usually adopted to reduce as much as possible the test duration, such as performing calculations while acquiring data: this would result in an almost unreadable SW but rather a high performance one. In our example we consider a straightforward programming approach, more helpful for understanding.

4.2.1   Setup and calibration.   The first task to be carried out by a test programme during loading, is to properly set all the hardware in a safe and known condition and to calibrate all system resources and test board hardware. Calibration of the test hardware is recommended especially for high accuracy measurements. Once the programme is loaded and calibration arrays are stored, the programme main routine is run each time a new device has to be tested.

Usually the first measurement performed after the power on of the device is a continuity test (to verify the electrical integrity of the connections, without which successive measurements on that device are useless and the device has to be discarded) and current leakage test.

The following code script[1] illustrates a typical start-up procedure.

---

[1] Courtesy of LTX Corporation - Westwood, MA.

```
------------------------------------------------------------------------
procedure on_start
------------------------------------------------------------------------
--  This routine contains the top level calling sequence for all sections
--  of the programme.
--  See Calibrate_hw for use of debug_mode

local    float    :    ttime_calc
local    integer :    R = 1         -- 24  -- number of records
body

     LTX_start   -- Initialisation of the ATE system Hardware
     -- on first run
     first_run    =    false
     if first_run then
         Initial_setup
         Calibrate_hw
         Digicomp_checker(DATA_BUS)
         -- calibration of a system part dedicated to ADC testing
     end_if
--   MEASURE INPUT AND SUPPLY RAIL CURRENTS
     Meas_dc_static
     Self_bias_voltage
--   FUNCTIONAL TESTS --
     Init_servo_test -- initialisation of a dedicated System HW for ADC testing
      Disq_bins([1..5]) -- Binning of the device for the following test
      Functionality
          -- pure functionality at various frequency: no parameters measure
      Disq_bins([1..6])  -- Binning of the device for the following test
      Test_output_levels      -- digital pins output level measurement
      Delay_times
--   MEASURE ACCURACY --
     Measure_accuracy("0_+2v",300)  -- static ADC characterisation
     Measure_dynamic(21MHz,1v+VOD,1.5,R)
     -- ft = 21MHz, 100KHz  input sinewave da 0v a 2v
-- Input Vpeak = 1v+VOD, Input DC offset= 1.5V,
-- R = number of records
--   HISTOGRAM TEST ANALYSIS
     Histogram_test(0v,2v,float(Mr),R)
     --   procedure Histogram_test(T1,T255,M=number
     --   of samples,R=number of records)
--   SINE FIT ANALYSIS
     SineFit_test
--   FFT TEST ANALYSIS
     Frequency_test
--   END OF TESTS
     if first_run then
         End_first_run
     end_if

end_body
```

After these tests, more specific performance measurements are executed; in our case the CH, SA and SF methods are implemented.

4.2.2   **Histogram method.**   Herein follows an example of implementation of the code
histogram analysis method .

```
--------------------------------------------------------------------------
procedure Histogram_test(T1,T255,M,R)
--------------------------------------------------------------------------
in float     :   T1 -- measured T1 transition level
in float     :   T255 -- measured T255 transition level
in float     :   M -- number of samples in each record
in integer   :   R --number of records
local
    float    :   S -- total number of samples for cumulated histogram
    float    :   kmax    ,kmin -- max and min value for histogram bins
    integer  :   k,i
    set[2]     : maxmin
    float    :   A,C -- input sinewave Amplitude and Offset measured
    float    :   e[NCODES] -- error array
    float    :   x[4] -- temporary variables where sums are stored
    float    :   G,Vos,INL,DNL,INLar[NCODES],DNLar[NCODES] -- parameters
                 -- to be measured
    float    :   Tnom[NCODES],Q ,Vfs=2.  -- nominal values
    float    :   Hc[NCODES] -- cumulative Histogram array
end_local

body

  S = M*float(R)
  Tnom = 0.
  T = 0.
  Hc = 0.
  -- NOMINAL VALUES CALCULATION
  for k=1 to NCODES-1 do
    Tnom[k] = (float(k)-0.5)*Vfs/float(2^NBITS)
    for i=1 to k do
      Hc[k] = Hc[k] + H[i] -- H array generated during acquisition
    end_for
  end_for
  Q = (Tnom[255] - Tnom[1]) / (float(2^NBITS) -2.)

  -- TRANSFER FUNCTION PARAMETERS CALCULATION
  A = (T255 - T1) / ( cos(PI*Hc[1]/S) + cos(PI*(1.-( Hc[NCODES-1]/S ))) )
  C = (T255*cos(PI*Hc[1]/S) + T1*cos(PI*(1.-(Hc[NCODES-1]/S)))) /
  (cos(PI*Hc[1]/S) + cos(PI*(1.- (Hc[NCODES-1]/S))))

  for k=1 to NCODES-1 do
    T[k] = C - A* cos(PI*Hc[k]/S) -- eq. 4.14
  end_for

  -- BEST FIT METHOD
  -- GAIN AND OFFSET ERRORS CALCULATION
  Best_fit_G_Vos(8,T,Tnom,Q,G,Vos)
  INL_DNL_tests_results(8,"best-fit",G,Vos,Tnom,Q)
  -- END POINT METHOD
  -- GAIN AND OFFSET ERRORS CALCULATION
  End_point_G_Vos(8,T,Tnom,G,Vos)
  INL_DNL_tests_results(8,"end-point",G,Vos,Tnom,Q)

end_body
```

```
------------------------------------------------------------------------
procedure Best_fit_G_Vos(NB,T,Tnom,Q,G,Vos)
------------------------------------------------------------------------
in  float   :   T[?],Tnom[?] -- theoretical and non theoretical arrays
in float    :   Q  -- LSB
in integer  :   NB
out float   :   G,Vos  -- calculated gain and offset errors

local
    float   :   sTk,sT,sT2  -- temporary variables where sums are stored
    integer :   k
    integer :   NCD
end_local

body

  NCD = 2^NB
  sTk = 0.0
  sT  = 0.0
  sT2 = 0.0
  for k=1 to NCD-1 do
    sTk = sTk + float(k)*T[k]
    sT  = sT  + T[k]
    sT2 = sT2 + T[k]*T[k]
  end_for

  G = Q* float(NCD-1)*(sTk - 2.^float(NB-1)*sT)/(float(NCD-1)*sT2-sT*sT)
  Vos = Tnom[1] + Q*( 2.^float(NB-1) -1.) - G*sT/float(NCD-1)

end_body

procedure End_point_G_Vos(NB,T,Tnom,G,Vos)
------------------------------------------------------------------------
in  float   :   T[?],Tnom[?]
in integer  :   NB
out float   :   G,Vos  -- gain and offset errors

body

  G = (Tnom[255]-Tnom[1]) / (T[255]-T[1])
  Vos = (Tnom[255]*T[1] - Tnom[1]*T[255]) / (T[1]-T[255])

end_body
```

In the following subroutine the INL and DNL parameters are evaluated and compared to the specification limits, printed to the data-log file and to the output and binning of the device is calculated.

```
-----------------------------------------------------------------------
procedure INL_DNL_tests_results(NB,method,G,Vos,Tnom,Q)
-----------------------------------------------------------------------
in string[10]  :   method   -- method: best-fit or end-point
in float    :   G    -- gain error parameter
in float    :   Vos   -- Offset error parameter
in float    :   Tnom[?]   -- nominal values
in float    :   Q    -- lsb or ideal quantisation level
in integer  :   NB    -- number of bits of the converter

local
    integer :   k
    float   :   e[NCODES]    -- error array
    float   :   INLar[NCODES],INL --INL array error and INL maximum error
    float   :   DNLar[NCODES],DNL --DNL array error and DNL maximum error
    float   :   x[4]
    string[20]  :   tnameI,tnameD
end_local

body

  Disq_bins([1..8])
  -- INL calculation
  for k=1 to NCODES-1 do
    e[k] = Tnom[k] - G*T[k] - Vos
    INLar[k] = e[k] / Q
  end_for
  x = xtrm(INLar)
  if abs(x[1]) > abs(x[3]) then
    INL = abs(x[1])
  else
    INL = abs(x[3])
  end_if
  tnameI = method+"_INL"
  test tnameI
    if INL > 1.01lsb then
        Fail
    end_if
    Bin_device
    dlog(tnameI,"@t",INL!u=LSB,"@t","lim= 1LSB")
  end_test
  -- DNL calculation
  DNLar = 0.0
  for k=1 to NCODES-2 do
    DNLar[k] = G*(T[k+1] - T[k])/Q - 1.
  end_for
  x = xtrm(DNLar)
  if abs(x[1]) > abs(x[3]) then
    DNL = x[1]
  else
    DNL = x[3]
  end_if
  tnameD = method+"_DNL"
  test tnameD
    if abs(DNL) > 0.51lsb then
```

```
        Fail
    end_if
    Bin_device
    dlog(tnameD,"@t",DNL!u=LSB,"@t","lim= 0.5LSB")
  end_test

end_body
```

4.2.3   **Spectral method.**   Here follows an example of implementation of the SA algorithm

.

```
---------------------------------------------------------------------
procedure Frequency_test
---------------------------------------------------------------------
local
   boolean : coherent
   integer : epsj10,lmax
   double  : corr_factor[10] -- correction factor for the
             -- first 10 harmonics
   double  : ENBW
end_local

body

   Coherence_check(epsj10,coherent)
   -- epsj10 is calculated and coherence is verified
   -- Order_it_back
   Ydb = 0.
   Ydb[1:(Mr/2 +1)] = mag_fft(Y[1:Mr])

   Find_real_principal_component(Jc)
   -- Jc = Jcalculated finding the maximum of the spectrum
   -- coherent = false

   -- Window, chosen by the operator interactively,
   -- is applied if coherence is not met
   Windowing(epsj10,coherent,lmax,corr_factor,ENBW)
   YFFT = 0.
   YFFT[1:(Mr/2 +1)] = mag_fft(Y[1:Mr])
   -- apply fft to windowed time domain array

   Frequency_parameters_tests(coherent,epsj10,lmax,corr_factor,ENBW)

end_body
```

```
procedure Windowing(epsj10,coherent,lmax,corr_factor,ENBW)
------------------------------------------------------------------------
in  integer :   epsj10
in  boolean :   coherent
out integer :   lmax
out double  :   corr_factor[10]
out double  :   ENBW

local
    integer : Wchosen
    string[10]  : answer
end_local

body

  if coherent then
    lmax = 0
    corr_factor = 1.
    ENBW = 1.
    Y = Y
  else
    println(stdout,"Non-coherent condition: suggested window for 8-bit
    ADC is 7terms Blackman-Harris")
    println(stdout,"Do you want to use another window?")
    input(stdin,answer)
    if answer = "yes" or answer="YES" or answer="Yes" or answer="si"
    or answer="SI" then
        println(stdout,"Non-coherent condition: choose a window
    to apply")
        println(stdout,"1 - Hanning")
        println(stdout,"2 - Hamming")
        println(stdout,"3 - Blackman")
        println(stdout,"4 - Exact Blackman")
        println(stdout,"5 - 7terms Blackman-Harris")
        input(stdin,Wchosen)
        if Wchosen = 1 then
            Apply_Hanning(epsj10,lmax,corr_factor,ENBW)
        else_if Wchosen = 2 then
            Apply_Hamming(epsj10,lmax,corr_factor,ENBW)
        else_if Wchosen = 3 then
            Apply_Blackman(epsj10,lmax,corr_factor,ENBW)
        else_if Wchosen = 4 then
             Apply_Exact_Blackman(epsj10,lmax,corr_factor,ENBW)
        else_if Wchosen = 5 then
            Apply_BH7(epsj10,lmax,corr_factor,ENBW)
        end_if
    else
        Apply_BH7(epsj10,lmax,corr_factor,ENBW)
    end_if
  end_if

end_body
```

```
procedure Apply_Hanning(epsj10,lmax,corr_factor,ENBW)
-------------------------------------------------------------------
in  integer :   epsj10  -- number of cycles uncertainty (uncertainty on J)
--  rounded to first decimal and multiplied by 10
out integer :   lmax  -- maximum side-bins to be excluded in parameters'
--  computations
out double  :   corr_factor[10]  -- correction factor for the first 10
--  harmonics
out double  :   ENBW  -- Equivalent Noise Bandwidth of the window

local
    double  :   SUMW,SUMW2
    double  :   wL[10*32768],W[10*32768]
    integer :   i,signal_bin,Ji
end_local

body

  SUMW = 0.
  SUMW2 = 0.

  for i=1 to Mr do
    wL[i] = 0.5 - 0.5*cos(2.*PI*double(i/Mr))
    Y[i] = Y[i]*wL[i]
    SUMW = SUMW + wL[i]
    SUMW2 = SUMW2 + (wL[i]*wL[i])
  end_for

  ENBW = double(Mr)*SUMW2 / (SUMW*SUMW)

  lmax = 2

  wL[Mr+1:10*Mr] = 0.
  W =0.
  Ji = round(Jc)
  signal_bin = 10*Ji+ epsj10

  dft(W[1:10*Mr], wL[1:10*Mr], signal_bin, integer(Fs), 10)

  for i = 1 to 10 do
    corr_factor[i] = W[3*i + 1]
  end_for

end_body
```

```
procedure Apply_Hamming(epsj10,lmax,corr_factor,ENBW)
----------------------------------------------------------------------
in  integer :   epsj10  -- number of cycles uncertainty (uncertainty on J)
--  rounded to first decimal and multiplied by 10
out integer :   lmax  -- maximum side-bins to be excluded in parameters'
-- computations
out double  : corr_factor[10]   -- correction factor for the first
--  10 harmonics
out double  :   ENBW  -- Equivalent Noise Bandwidth of the window

local
    double  :   SUMW,SUMW2
    double  :   wL[10*32768],W[33]
    integer :   i,signal_bin,Ji
end_local

body

  SUMW =0.
  SUMW2 = 0.

  for i=1 to Mr do
    wL[i] = 0.54 - 0.46*cos(2.*PI*double(i/Mr))
    Y[i] = Y[i]*wL[i]
    SUMW = SUMW + wL[i]
    SUMW2 = SUMW2 + (wL[i]*wL[i])
  end_for

  ENBW = double(Mr)*SUMW2 / (SUMW*SUMW)

  lmax = 2

  wL[Mr+1:10*Mr] = 0.
  W = 0.
  Ji = round(Jc)
  signal_bin = 10*Ji+ epsj10

  dft(W[1:10*Mr], wL[1:10*Mr], signal_bin, integer(Fs), 10)

  for i = 1 to 10 do
    corr_factor[i] = W[3*i + 1]
  end_for

end_body
```

```
procedure Apply_Blackman(epsj10,lmax,corr_factor,ENBW)
----------------------------------------------------------------------
in   integer :   epsj10  -- number of cycles uncertainty (uncertainty on J)
--   rounded to first decimal and multiplied by 10
out integer :   lmax  -- maximum side-bins to be excluded in parameters'
--   computations
out double  :   corr_factor[10]  -- correction factor for the first
--   10 harmonics
out double  :   ENBW  -- Equivalent Noise Bandwidth of the window

local
    double  :   SUMW,SUMW2
    double  :   wL[10*32768],W[33]
    integer :   i,signal_bin,Ji
end_local

body

  SUMW =0.
  SUMW2 = 0.

  for i=1 to Mr do
    wL[i] = 0.42-0.5*cos(2.*PI*double(i/Mr))+0.08*cos(4.*PI*double(i/Mr))
    Y[i] = Y[i]*wL[i]
    SUMW = SUMW + wL[i]
    SUMW2 = SUMW2 + (wL[i]*wL[i])
  end_for

  ENBW = double(Mr)*SUMW2 / (SUMW*SUMW)

  lmax = 3

  W=0.
  wL[Mr+1:10*Mr] = 0.
  Ji = round(Jc)
  signal_bin = 10*Ji+ epsj10

  dft(W, wL[1:10*Mr], signal_bin, integer(Fs), 10)

  for i = 1 to 10 do
    corr_factor[i] = W[3*i + 1]
  end_for

end_body
```

```
procedure Apply_BH7(epsj10,lmax,corr_factor,ENBW)
-----------------------------------------------------------------------
in  integer :   epsj10  -- number of cycles uncertainty (uncertainty on J)
--  rounded to first decimal and multiplied by 10
out integer :   lmax  -- maximum side-bins to be excluded in parameters'
--  computations
out double  :   corr_factor[10]  -- correction factor for the first
--  10 harmonics
out double  :   ENBW  -- Equivalent Noise Bandwidth of the window

local
    double  :   SUMW,SUMW2
    double  :   wL[10*32768],W[33]  -- 10harmonics *3 + 3
    double  :   a0,a1,a2,a3,a4,a5,a6  -- window coefficients
    integer :   i,signal_bin1,signal_bin2,Ji
end_local

body

  SUMW =0.
  SUMW2 = 0.
  a0 = 0.271051400693424
  a1 = 0.433297939234485
  a2 = 0.218122999543110
  a3 = 0.065925446388031
  a4 = 0.010811742098371
  a5 = 0.000776584825226
  a6 = 0.000013887217352
  for i=1 to Mr do
  wL[i] = a0-a1*cos(2.*PI*double(i/Mr))+a2*cos(4.*PI*double(i/Mr))-
  + a3*cos(6.*PI*double(i/Mr))+ a4*cos(8.*PI*double(i/Mr))-
  + a5*cos(10.*PI*double(i/Mr))+a6*cos(12.*PI*double(i/Mr))
    Y[i] = Y[i]*wL[i]
    SUMW = SUMW + wL[i]
    SUMW2 = SUMW2 + (wL[i]*wL[i])
  end_for

  ENBW = double(Mr)*SUMW2 / (SUMW*SUMW)

  lmax = 7
  W=0.
  wL[Mr+1:10*Mr] = 0.
  Ji = round(Jc)
  signal_bin1 = 10*Ji+ epsj10
  signal_bin2 = 10*Ji - epsj10
  dft(W, wL, epsj10, (10*Mr),10)
  corr_factor[1] = SUMW / W[4]
  dft(W, wL, signal_bin1, (10*Mr),10)
  for i = 2 to 10 do
    corr_factor[i] = SUMW / W[3*i + 1]
  end_for

end_body
```

```
procedure Coherence_check(epsj10,coherent)
----------------------------------------------------------------------
-- Here epsj10 is the uncertainty in J (see eq. 6.17)
-- j calculated only for multiple of 0.1 and
-- multiplied by 10, thus having epsj10 an integer =[1,2,3,4,5];
-- this was done following Subsection -- 6.9.1 indications

out    integer  : epsj10
out boolean    : coherent

local
    double  :   epsj
    double  :   x
end_local

body

  x = double(Mr)*double(Fi)/double(Fs)
  J = round(x)
  epsj = x - double(J)
  epsj10 = integer(10.*epsj)

  if epsj10 < 1 then
    coherent = true
    epsj10 = 0
  else
    coherent = false
  end_if

end_body

procedure Find_real_principal_component(Jc)
----------------------------------------------------------------------
out float  :   Jc

local
    double  :  max_min[4]
    integer :   i
end_local

body

  --Ydb[1] = 0.  --set dc-component to zero
  mag_to_dbc(Ydb[1:(Mr/2+1)],Ydb[1:(Mr/2+1)])
  --Ydb = power_ratio_to_db(Ydb)  -- in dBm  (dB mW)
  Ydb[1] = -120.
  max_min = xtrm(Ydb[1:Mr/2])

  Jc = float(max_min[2]) - 1.  --position of the maximum in the array

end_body
```

Calculation of the various parameters.

```
procedure Frequency_parameters_tests(coherent,epsjr,lmax,corr_factor,ENBW)
-----------------------------------------------------------------------------
in boolean  :  coherent
in integer  :  epsjr  -- epsilon_j (uncertainty on frequency ratio) rounded to
--  first decimal and multiplyed by 10
in integer  :  lmax  -- maximum number of sidelobes to be excluded in
--  parameter's calculation
in double   :  corr_factor[10]  -- correction factor to be applied to
--  harmonics' components amplitude
in double   :  ENBW  -- Equivalent Noise Bandwidth of
--  the chosen window

local
    double  : SNR,SINAD,NF1,NF12,NF1db,NFdbc,NFdbFS,SUMQ,THD,Nef,SFDR,SFSR
    -- parameters to be evaluated
    double  : dSUMQ
    float   : epsj
    integer : k,h,l,Ji,f_h_sp,i1,i2
    double  : a,b,d,c,e,sf  -- intermediate calculation parameters
    integer : Jh[20]
    integer : Mfft_max
end_local
const h_max = 10  -- maximum number of harmonics to be considered

body
  epsj = float(epsjr) / 10.
  set NF1 = 0. SUMQ = 0. SNR = 0. THD = 0. SINAD = 0. SFDR = 0. dSUMQ = 0. Jh = 0
  Ji = integer(Jc)
  Mfft_max = Mr/2 + 1
  Find_h_indexes(Ji,Mr,Jh)
  Ji = integer(Jc) + 1
  SUMQ_calculation(coherent,h_max,Jh,lmax,epsj,SUMQ)

test "Frequency An NF1"
    NF12 = (SUMQ + 0.5*YFFT[Mfft_max]*YFFT[Mfft_max]) /
    ( double(Mr/2) - (double(h_max)*(2.*double(lmax) + 1.)) )
    NF1 = sqr(NF12)
    NFdbc = 10.*log(NF12/(YFFT[Ji]*YFFT[Ji]))
    Bin_device
    dlog("NF1= ",NFdbc!u=dBc)
end_test

test "Frequency An SNR"
    c = 10.*log( abs((YFFT[Ji]*YFFT[Ji]) - NF12) / NF12 )
    e = c - 10.*log( double(Mr)/2. + 1.)
    SNR = e + 10.*log(ENBW) + 10.*log(corr_factor[1]*corr_factor[1])
    Bin_device
    dlog("Spectral analysis","@t","SNR @t",SNR!u=dB)
end_test

test "Frequency An THD"
    d = 0.
    for h=2 to h_max do
        d = d + YFFT[Jh[h]]*YFFT[Jh[h]]*corr_factor[h]*corr_factor[h]
    end_for
    THD = 10.*log( d/(YFFT[Ji]*YFFT[Ji]*corr_factor[1]*corr_factor[1]) )
```

```
        Bin_device
        dlog("Spectral analysis","@t","THD @t",THD!u=dB)
    end_test

    test "Frequency An SFDR"
        Calculate_max_non_principal_spectral_component(Jh,h_max,f_h_sp)
        SFDR = 10.*log( YFFT[Ji]*YFFT[Ji] / (YFFT[f_h_sp]*YFFT[f_h_sp]) ) +
        + 10.*log( corr_factor[1]*corr_factor[1] )
        Bin_device
        dlog("Spectral analysis","@t","SFDR @t",SFDR!u=dB)
    end_test

    test "Frequency An SINAD"
        b = 0.
        if coherent then
          a = 2.*NF12 + 0.5*YFFT[(Mfft_max)]*YFFT[(Mfft_max)]
          for k = 2 to (Mfft_max-1) do
              if k <> Ji then
                a = a + YFFT[k]*YFFT[k]
              end_if
          end_for
        else
          a = (2.*double(lmax) + 2.)*NF12 + 0.5*YFFT[(Mr/2+1)]*YFFT[(Mr/2+1)] + SUMQ
          for h=2 to h_max do
              i1 = Jh[h] + round(float(h)*epsj)
              i2 = Jh[h] - round(float(h)*epsj)
              if i1 < Mr/2 then
                b = b + ENBW * (YFFT[i1] * YFFT[i1]) * (corr_factor[h]*corr_factor[h])
              end_if
              if i2 < Mr/2 then
                b = b + ENBW * (YFFT[i2] * YFFT[i2]) * (corr_factor[h]*corr_factor[h])
              end_if
          end_for
        end_if
        SINAD = 10.*log( (YFFT[Jh[1]]*YFFT[Jh[1]] - NF12)/(a+b) ) + 10.*log(ENBW) +
        + 10.*log( corr_factor[1]*corr_factor[1] )
        Bin_device
        dlog("Spectral analysis","@t","SINAD @t",SINAD!u=dB)
    end_test

end_body
```

```
procedure SUMQ_calculation(coherent,hmax,Jh,lmax,ej,SUMQc)
-----------------------------------------------------------------------------
in   boolean :    coherent
in   integer :    hmax,Jh[?],lmax
in   float   :    ej
out  double  :    SUMQc

local
     double  :    spectrum[32769]
     integer :    h,k,l,i1,i2
end_local

body
  spectrum = 0.
  SUMQc = 0.
  spectrum[1:Mr/2+1] = YFFT[1:Mr/2+1]

  spectrum[1] = 0.
  if coherent then
    for h=1 to hmax do
        spectrum[Jh[h]] = 0.
    end_for
  else
    for l = 0 to lmax do
        spectrum[(Jh[1]) + l] = 0.
        spectrum[(Jh[1]) - l] = 0.
    end_for
    for h = 2 to hmax do
        i1 = round(float(Jh[h]) - float(h)*ej)
        i2 = round(float(Jh[h]) + float(h)*ej)
           for l = 0 to lmax do
               spectrum[i1 + l] = 0.
               spectrum[i2 - l] = 0.
           end_for
    end_for
  end_if
  for k = 2 to (Mr/2) do
    SUMQc = SUMQc + (spectrum[k]*spectrum[k])
  end_for

end_body
```

```
procedure Calculate_max_non_principal_spectral_component(Jh,hmax,f_h_sp)
--------------------------------------------------------------------------------
in  integer :   Jh[20],hmax
out integer :   f_h_sp -- position of the maximum spectral component present
--  in the spectrum (except main one and DC)

local
    double  :   max_min[4],fftarray[32769]
    integer :   i
end_local

body

fftarray = YFFT
fftarray[1] = 0.  -- set dc component to zero level
fftarray[Jh[1]] = 0.  -- set principal component to zero level

--mag_to_dbc(fftarray[1:Mr/2+1],fftarray[1:Mr/2+1])

max_min = xtrm(fftarray)
f_h_sp = integer(max_min[2])  -- found position of maximum spectral component

end_body

procedure Find_h_indexes(Ji,Mr,Jh)
--------------------------------------------------------------------------------
-- Find position of the harmonics in the FFT array

in  integer :   Ji   -- fundamental tone index
in  integer :   Mr   -- Mr/2 maximum spectrum length
out integer :   Jh[20]

local
    integer :   h,Jmax[20]
    double  :   xtrm_arr[4],pfft[32769]
end_local

body

pfft = Ydb
Jh = 0
Jmax = 0
pfft[1] = -100.

for h = 1 to 20 do
    Jh[h] = Nyquist(h*Ji,Mr) + 1
    -- find position of the harmonics in Nyquist Band
    xtrm_arr=xtrm(pfft[1:Mr/2+1])
    Jmax[h] = integer(xtrm_arr[2])
    pfft[Jmax[h]] = xtrm_arr[3]
end_for

end_body
```

```
function Nyquist(Jj,Jfs) :    integer
--------------------------------------------------------------------------------
in  integer :   Jj   -- bin of tone in spectrum array
in integer  :   Jfs  -- bin of sampling frequency in spectrum array

local
    integer :   Jh   -- bin of harmonic component
    integer :   nj
end_local

body

if Jj <= Jfs/2 then
    return(Jj)
else
    nj = 1
    Jh = abs(Jj - nj*Jfs)
    while Jh > Jfs/2 do
        nj = nj+1
        Jh = abs(Jj - nj*Jfs)
    end_while
        return(Jh)
end_if

end_body
```

4.2.4 **Sine Fitting method.** Concerning the sine fitting method, only the known frequency ratio case was considered because, when using this particular ATE, one may operate and calibrate it in such a way to be in that condition .

```
procedure SineFit_test
--------------------------------------------------------------------------------
-- NB record_data array corresponds to Y array in DYNAD draft (output ADC data)
--------------------------------------------------------------------------------


local

double   :    xp[3]
-- input parameter vector xp[1]=Acos(teta) , xp[2]=Asin(teta), xp[3]=C
double   :    x[65536] --[32768]
-- discrete points of x(t) sampled at Fs x[n] = A*cos(omega_i*n + teta) + C
double   :    dxy[65536]
double   :    A , C , teta  -- input signal amplitude, offset and initial phase
double   :    omega_i  -- omega_i = 2*PI*Fin/Fs
double   :    x_rms,eta_rms  -- intermediate calculation parameters
double   :    x2_rms,eta2_rms,diff
double   :    SINAD
integer  :    i,n
float    :    t_sinefit  -- test time for sinefit algorithmic calculation
double   :    jf

end_local

body

start_timer

-- analytical solution for the case of known frequency ratio
jf = double(Fi)* double(Mr) / double(Fs)
J = round(jf)
omega_i = 2.*PI* double(J)/double(Mr)
xp = 0.0
x = 0.
for n = 1 to Mr do
    xp[1] = ( 2.*Y[n]*cos(double(n)*omega_i) )/double(Mr) + xp[1]
    xp[2] = (-2.*Y[n]*sin(double(n)*omega_i) )/double(Mr) + xp[2]
    xp[3] = Y[n]/double(Mr) + xp[3]
end_for

-- initial parameter calculation: A,C,teta
teta = atn(xp[2]/xp[1])
A = xp[1] / (cos(teta))
C = xp[3]

-- x_rms and eta_rms calculation
x2_rms = 0.
eta2_rms = 2.
x = 0.
dxy = 0.
for n = 1 to Mr do
    x[n] = A*cos(omega_i*double(n) + teta) + C
    diff = double(x[n] - xp[3])
    x2_rms = diff*diff   + x2_rms
    dxy[n] = double(Y[n]) - x[n]
```

```
        eta2_rms = (double(Y[n] - x[n])*double(Y[n] - x[n])) + eta2_rms
end_for
x_rms = sqr(x2_rms/double(Mr))
eta_rms = sqr(eta2_rms / double(Mr))

test "Sine Fit SINAD"
    SINAD = x_rms / eta_rms
    SINAD = 20.*log(SINAD)    -- in dB
    Bin_device
    dlog("Sine Fit @t","@t", "SINAD @t",SINAD!u=dB)
    println(stdout,"Sine Fit SINAD= ",SINAD!u=dB)
end_test

end_body
```

4.3   Results

The results obtained are summarized in the final data-log file. However to analyse specific parameters it is possible to visualize them in a graphical form. Herein follows the representations of the main parameters of the histogram analysis obtained from the measurement of ADC devices. Figure 6.10 shows the histogram

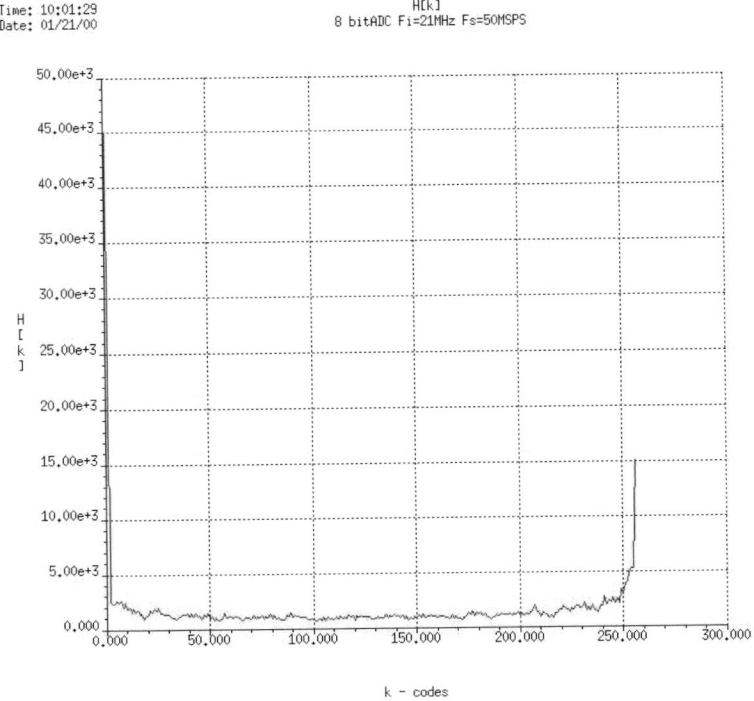

Time: 10:01:29
Date: 01/21/00

H[k]
8 bitADC Fi=21MHz Fs=50MSPS

Figure 6.10.   Histogram obtained from 24 records with a 16384 samples length.

$H[k]$ as a function of code bin k. Figure 6.11 illustrates the behaviour of $INL[k]$ and $DNL[k]$ as a function of code k, using the best fit definition. Figure 6.12 shows the behaviour of $INL[k]$ and $DNL[k]$ as a function of code k, using the end-points definition. The INL and DNL errors are defined to be the maximum in absolute value in the corresponding arrays.

The spectrum obtained by the SA method is reported in figure 6.13. The first ten harmonic components are indicated. All spectral parameters are calculated from this array.

Figure 6.14 shows the sinewave reconstructed after the acquisition, which can be compared to the theoretical input one. It is generally recommended to take a glance at the reconstructed waveform, to point out possible errors in the test setup: the figure shows a case, where the offset of the input sinewave was not properly chosen, and thus the waveform is clipped at code 0. The algorithm used to reconstruct a period of the sinewave by reordering the acquired samples is given in the following sub-routine

```
-- REORDERING FORMULA for sub-sampling case : frecord_data[] -> record_data_ord[]


NUM_PER = 6903    -- M according Matt. Mahoney (calculated as ft/fs * N)

body
     record_data_ord=0.
     for k=1 to M do
```

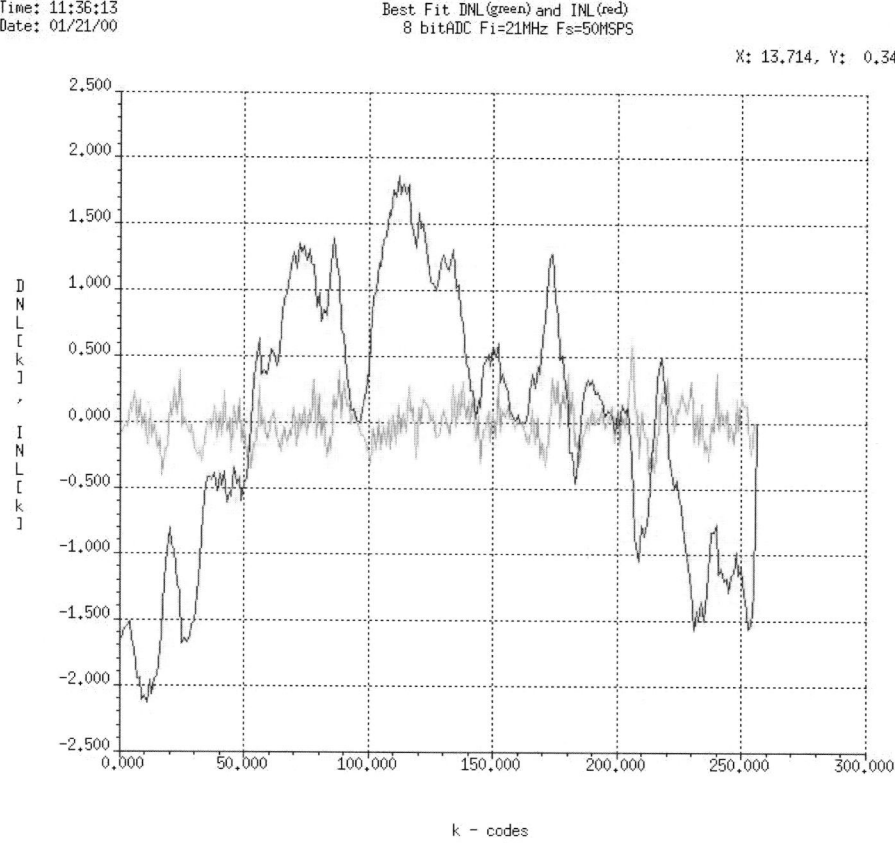

Time: 11:36:13
Date: 01/21/00

Best Fit DNL(green) and INL(red)
8 bitADC Fi=21MHz Fs=50MSPS

X: 13.714, Y: 0.346

Figure 6.11. Best fit INL and DNL arrays.

```
    record_data_ord[NUM_PER*(k-1)+1-M*integer(NUM_PER*(k-1)/M)] = frecord_data[k]
    end_for
    wait(0ms)          -- BREAK POINT TO SEE SINEWAVE REORDERED
end_body
```

Then, an example of output data-log for each device tested follows.

```
File record out_1:          (Input sinewave =< FSR)
Analog(5V)+Digital(5V)_current   7.64mA  max=36ma
Analog(5V)_(dig@3.3V)_current    7.60mA  max=33ma
Digital(3.3V)_(an@5V)_current    0.10mA  max=3ma
Sine Fit  SINAD   28.65dB
Spectral analysis SNR    30.85dB lim=38dB(fi=10MHz)
Spectral analysis THD   -32.68dB
Spectral analysis SFDR   34.31dB lim=45dB (fi=10MHz)
Spectral analysis SINAD  28.66dB

File record ovrd24_1b:  (Input sinewave FSR + Vod)
Analog(5V)+Digital(5V)_current   10.67mA  max=36ma
```

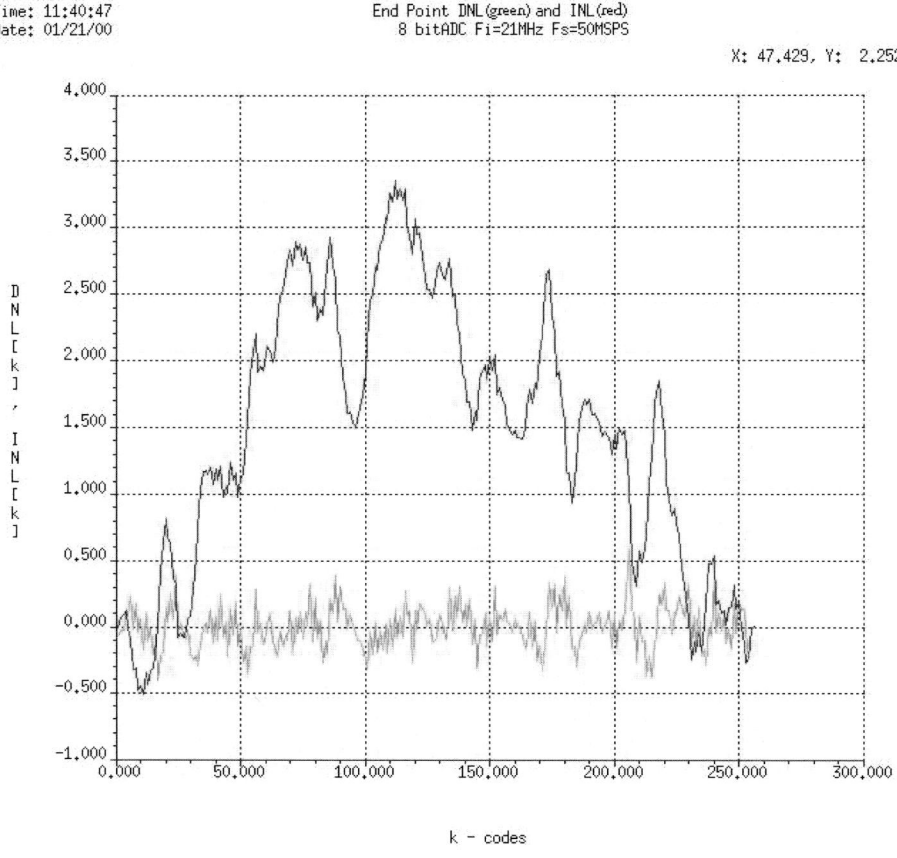

Figure 6.12.   End-points INL and DNL array.

```
Analog(5V)_(dig@3.3V)_current   10.62mA   max=33ma
Digital(3.3V)_(an@5V)_current 0.11mA   max=3ma
best-fit_INL    2.15LSB
best-fit_DNL    0.65LSB
end-point_INL   3.39LSB max=1.5LSB (fi=10MHz)
end-point_DNL   0.65LSB max=0.5LSB (fi=10MHz)
```

4.4   Comparison

The main comparison that can be made between the three proposed methods regards the ease of implementation and test duration. Once the equations are well understood, there is no particular difficulty in their implementation. The testing time of the three methods, instead, makes a difference. The acquisition of a single 16384 samples record at 50 MS/s takes 4.5 seconds within this ATE, using the internal Vector Processor for parallel reordering of the record (without the Vector Processor utility the test time for a 16384 record would be about 26s). It is worth noting that reordering is not necessary for the CH analysis. The elaboration phase duration, i.e., the time each proposed algorithm (CH, SF, and SA) takes to calculate all the parameters mentioned from a 16384 length record, are

- CH → 4s (transition levels, INL, DNL)

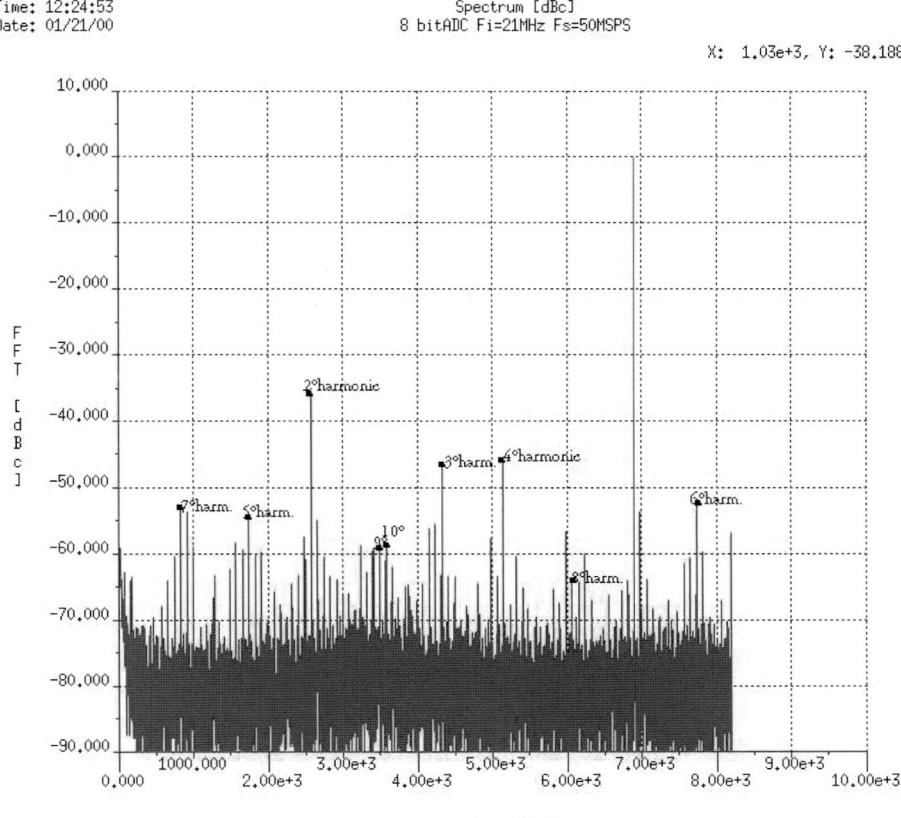

Figure 6.13. Spectrum graph.

- SA → 5s (SNR, THD, SFDR, SINAD)
- SF → 16s (SINAD)

The overall test time of the three methods is given by the acquisition time plus the elaboration time. Considering that for the CH analysis 24 records are necessary, we obtain then

- CH → 112s (transition levels,INL,DNL)
- SA → 9.5s (SNR,THD,SFDR,SINAD)
- SF → 20.5s (SINAD)

Thus, the fastest method is SA. However, if information on the converter linearity is needed, CH is the best choice, since INL and DNL cannot be directly calculated from spectrum analysis data. Finally, a distinction should be made between characterization testing (all codes and full analysis) and production testing where a reduced number of code transition levels are tested and record lengths are shorter.

5. Conclusions

In this chapter a comparison of the three classical sinewave dynamic ADC test methods — spectral analysis (SA), sinewave fitting (SF), and code histogram (CH) — is presented. This comparison is performed

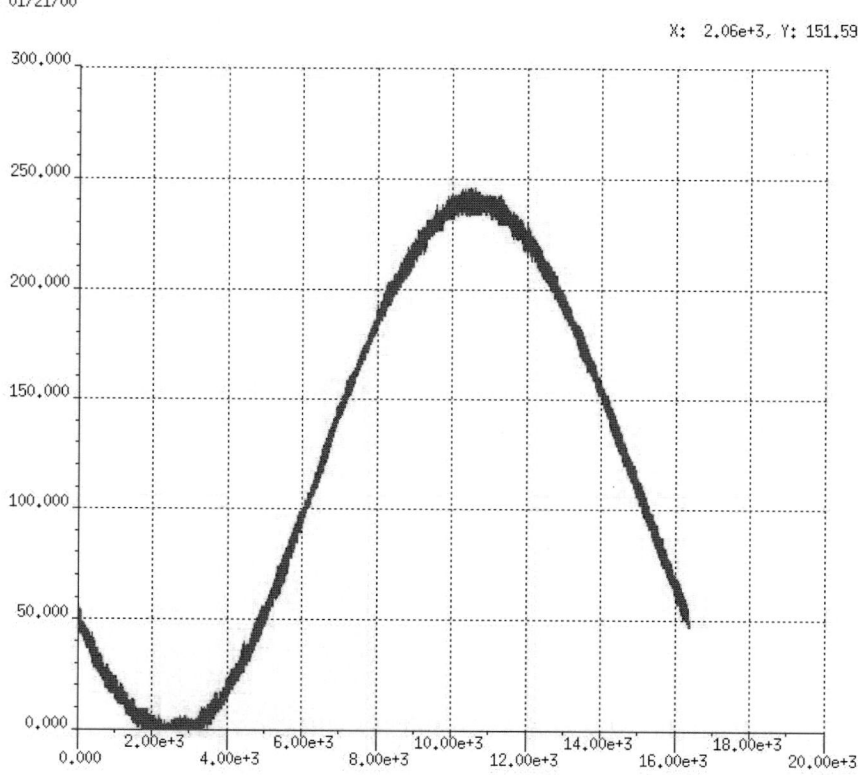

Time: 12:20:09
Date: 01/21/00

Figure 6.14.   Sine Fit: input sinewave from output data

in terms of the accuracy and precision obtained when estimating the Signal to Noise-and-Distortion ratio, a parameter that can be measured with each one of these methods. These are also compared in terms of the minimum number of samples required for a certain accuracy and precision, as well as in terms of their sensitivity to noise and distortion of the test stimulus, testing time, and of the sampling coherence.

The SA and SF methods are similar in terms of the results' accuracy they provide, the major differences being the larger data processing time required by the SF method and the necessity for windowing required by the SA one, in case coherent sampling can not be guaranteed. The main advantages of the SA method over the other two are the shortest testing time and largest set of parameters it provides. The SF method is more robust than SA to non-coherency, but the limited number of parameters that can be extracted might be a handicap for its application.

The CH method requires a much larger number of samples than the other two, and thus, although its data processing time is the shortest, the resulting total test time is the longest. However, it seems to be the most robust when noise and harmonic distortion is present in the input stimulus, and allows characterising accurately the transfer characteristic of the ADC which is a critical requirement for certain applications. In fact, besides the advantages and disadvantages presented by each one of the methods, the final applications may dictate which method (or methods) should be used. This selection is often a balance between the choice of parameters required to be evaluated and the total allowed testing time.

An example of implementation on an ATE of these three algorithms was given together with the results obtained for a commercial 8bit 50 MS/s ADC. Considerations were made on mass-production criteria for the choice of the test method.

II

MEASUREMENT OF ADDITIONAL PARAMETERS

Chapter 7

JITTER MEASUREMENT

Pierre-Yves Roy
now with
EADS Defence and Security Systems SA
Defence and Communications Systems
Rue Jean-Pierre Timbaud - Montigny le Bretonneux
78063 Saint Quentin Yvelines Cedex, France
pierre-yves.roy@eads-telecom.com

Jacques Durand
now retired from
THALES
L'Orée de Corbeville, BP 56 91401 Orsay, France

1.   Introduction

For many new architectures of signal receivers (digital receivers) the analogue to digital conversion is performed on a carrier. An ideal digital receiver
would use a high dynamic range ADC at the output of the antenna, which
means that the ADC would digitise very high frequency signals (for instance
900MHz or 1800MHz for the European standard GSM). Today, this ideal digital receiver is not feasible and analogue down-conversion is still performed
because of the performances of the ADCs.

One of the limiting factors is the degradation of the ADC performances
as the input frequency increases. The degradation can be classified in two
categories. The first one is a decrease of the SFDR of the ADC and the second
one is a decrease of the SNR. The decrease of the SNR as the input frequency
increases, can be predicted if the jitter of the ADC is accurately measured.
That is why, for the state of the art signal receivers, the accurate measurement
of ADC's jitter is necessary in order to define the best architecture for the
receiver.

For the state of the art high speed ADCs, the traditional techniques [10],
[130] are note accurate enough to measure jitter precisely . Two methods based

D. Dallet and J. Machado da Silva, (eds.), Dynamic Characterisation of
Analogue-to-Digital Converters, 219–233.
© 2005 *Springer. Printed in the Netherlands.*

on a dual-measurement system overcome the problems of the additive noise of the test setup and of the linearity of the ADC that limited the traditional techniques. These two methods are described in this chapter. The first one referred as "the double beat technique" is detailed in [88] and the second one, "the joint probability technique" is taken from [33] and [34].

The last measurements performed with the double beat technique on a 14-bit 65 MS/s ADC are also described.

2.   The double beat technique

2.1   Test setup and principle

The dual-channel setup used is depicted in figure 7.1. Two phase-locked synthesizers are used to generate the clock and input signal. The input sine waves as well as the clock signals are the same for the two ADCs. Band-pass filters are used to reduce the noise generated by the synthesizers.

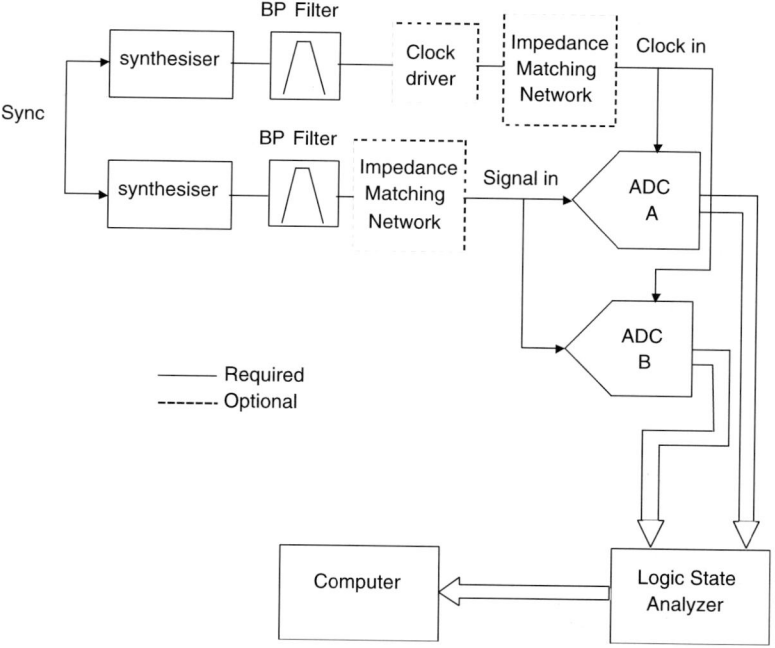

Figure 7.1.   Test setup for the double beat technique.

The technique described in this chapter is based on a signal processing procedure that can be divided into three successive steps:

1   The double beat to eliminate systematic errors

2   The subtraction to eliminate synthesizer's phase noise

3 The extraction to remove the ADCs' amplitude noise and to demodulate the noise by the slope of the input sine wave

The different steps and the resulting signals are described in figure 7.2.

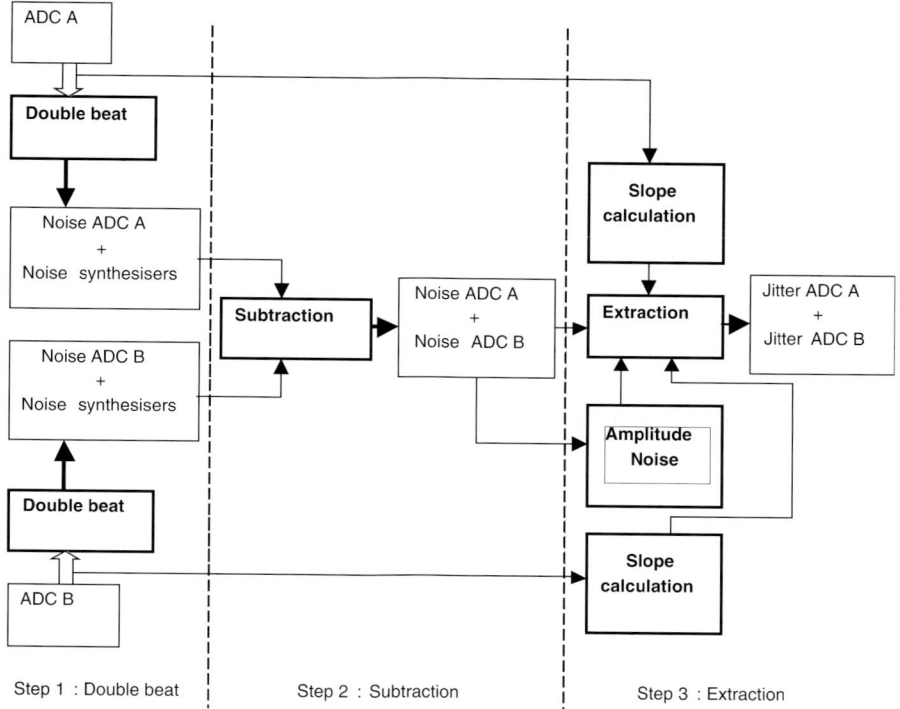

Figure 7.2. Principle of the subtraction technique.

The processing described above leads to the sum of the jitters of ADCs A and B. To calculate the jitter of each ADC (σ_{jA}^2 and σ_{jB}^2), a third ADC (ADC C) is needed in order to perform three successive identical measurements. The first one performed with ADCs A and B gives $\sigma_{j1}\sqrt{\sigma_{jA}^2 + \sigma_{jB}^2}$, the second one with ADCs A and C gives $\sigma_{j2}\sqrt{\sigma_{jA}^2 + \sigma_{jC}^2}$, and finally the last measurement with ADCs B and C gives $\sigma_{j3}\sqrt{\sigma_{jB}^2 + \sigma_{jC}^2}$. Then, the jitter of each ADC is calculated by combining σ_{j1}, σ_{j2}, and σ_{j3}.

2.2 The double beat

As described in figure 7.3, the ADCs under-sample the input sine wave in order to get at the output of each ADC a beat sine wave whose period is half the acquisition duration .

To perform an accurate measurement, the input frequency must be chosen as high as possible. Moreover, to get two periods of the beat signal in the record length, the input frequency and the sampling frequency must satisfy:

$$f_{in} = (pM + 2)\frac{f_s}{M} \qquad (7.1)$$

where:

- f_{in} is the input frequency,

- f_s is the sampling frequency

- M is the number of samples,

- and p is and integer as large as possible.

The minimum number of samples to acquire is $\pi 2^N$, where N is the number of bits of the ADC. In practice, M is often set to 2^{N+2}.

The maximum value of p is derived from the maxim value of the input frequency given by the input bandwidth of the ADC under test. Once the acqui-

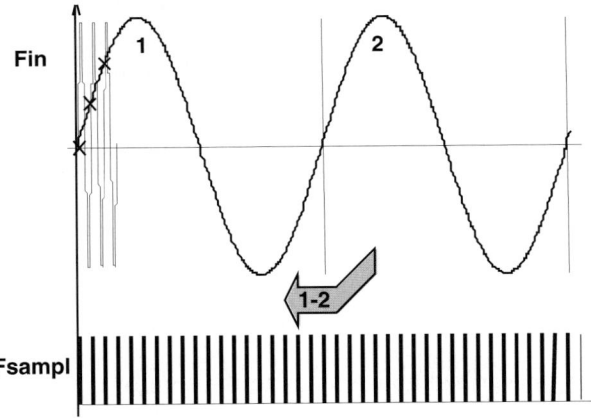

Figure 7.3.   Double beat principle.

sition is complete, the second period of the beat signal is subtracted point by point from the first one. This processing eliminates the signal and the systematic errors, such as non-linearities and quantization noise, and adds quadratically the uncorrelated noises. The resulting signal on each channel is depicted in figure 7.4. This signal is a noise composed of the ADC noise, as well as of the noise introduced by the synthesizers.

The noise represented above indicates from its shape that it is a jitter mainly induced noise. The superimposed amplitude noise is visible and measurable on the A-A' noise curve section which corresponds to the minima and maxima of the sinusoidal signal where the slope is equal to zero.

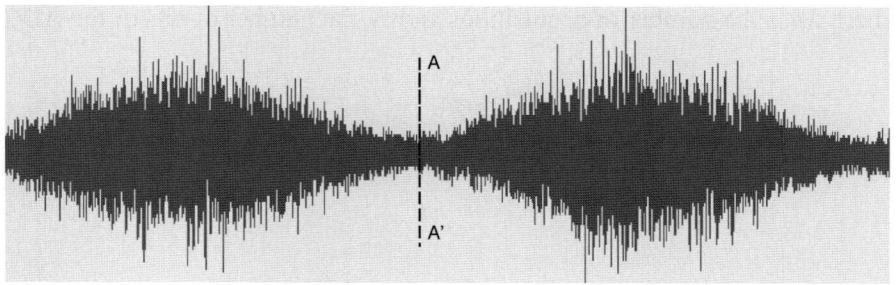

Figure 7.4. Resulting signal after the double beat processing.

2.3 The subtraction

The processing described in the former chapter is performed on the two channels. On each channel, the resulting signal is a noise composed by the ADC noise and by the noise of the synthesizers.

As both ADCs use the same input signal and the same clock signal, the noises due to the synthesizers on channels A and B are identical. Thus, the point to point subtraction of the signals on channels A and B removes the synthesizers' noises. As the noises due to the ADCs are uncorrelated, they are quadratically added.

After this step, the signal is the quadratic sum of the noises of ADCs A and B. Its shape is similar to the one of the signal depicted in figure 7.4.

2.4 The extraction

Two successive processing must be performed to get the value of the jitter (this value equals the quadratic sum of the jitters of ADCs A and B). The first step removes the amplitude noise of the ADCs, thus the resulting signal is composed only by the noise due to the jitter of the ADCs. In the second step, the slope of the input sinewave is accurately calculated and the result is used to demodulate the noise and hence calculate the standard deviation of the jitter.

Elimination of the amplitude noise

As explained previously, the shape of the noise at this point is similar to the one of the noise depicted in figure 7.4. As the noise on AA' corresponds to the zero slope of the sinewave, it is caused only by the amplitude noise of the ADCs since jitter would not induce any noise at that point. In order to calculate the rms value of the amplitude noise, the width of minimum slope must be determined (figure 7.5). The number of points on the step of minimum slope is:

$$\Delta_1 = 2k_1 = 2\frac{M}{4\pi}\arccos\left[\frac{2^{N-1}-1}{2^{N-1}}\right] \tag{7.2}$$

where M is the number of acquisitions and N the number of bits of the ADC.

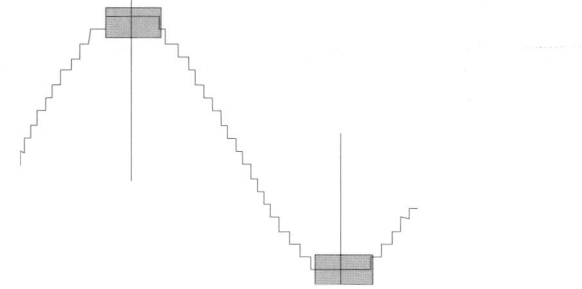

Figure 7.5. Width of amplitude noise measurement

In order to avoid introducing any kind of jitter, the amplitude noise is not calculated on the width Δ_1 but on a lower number of points (figure 7.6). The

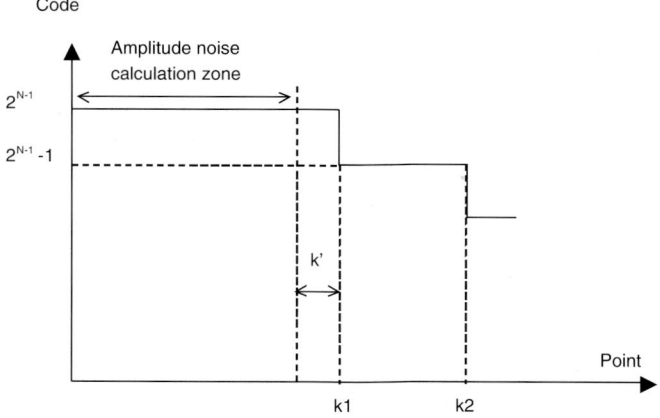

Figure 7.6. Exact width of amplitude noise calculation

number of points k' to remove from Δ_1 for the calculation of the amplitude noise depends on the jitter induced noise at point k_2 (the slope of the sinewave is considered constant between k_1 and k_2) and of the number of samples in code bin $2^{N-1} - 1$:

$$k' = [\textit{Number of samples in code } 2^{N-1} - 1] \times \textit{Proba[jitter induced noise at point } k_2 > 0.5LSB]$$

Indeed, if the jitter induced noise at point k_2 is greater than 0.5 LSB, some samples ideally quantized in code bin $2^{N-1} - 1$ can be quantized in code bin 2^{N-1} and these samples must not be taken into account for the calculation of the amplitude noise.

Number of samples in code bin $2^{N-1} - 1$

As k_2 is given by

$$k_2 = \frac{M}{4\pi} \arccos \left\{ \frac{2^{N-1} - 2}{2^{N-1}} \right\}$$ (7.3)

the number of samples in code bin $2^{N-1} - 1$ is given by

$$\Delta_2 = \frac{M}{2\pi} \left(\arccos \left\{ \frac{2^{N-1} - 2}{2^{N-1}} \right\} - \arccos \left\{ \frac{2^{N-1} - 1}{2^{N-1}} \right\} \right)$$ (7.4)

The jitter induced noise at point k_2

The slope of the input signal at point k_2 is

$$P_{k2} = 2^{N-1} \times 2\pi \frac{pM \pm 2}{M\sqrt{2}} \sin \left[\arccos \left(\frac{2^{N-1} - 1}{2^{N-1}} \right) \right]$$ (7.5)

The rms slope of the input signal is

$$P_{k2} = 2^{N-1} \times 2\pi \frac{pM \pm 2}{M\sqrt{2}}$$ (7.6)

From the rms value of the noise calculated after the subtraction step (σ_{tot}), it is possible to calculate the rms value of the noise induced at point k_2 by a pseudo jitter:

$$\sigma_{k2} = \frac{\sigma_{tot}}{\sqrt{2}} \frac{P_{k2}}{P_{rms}}$$ (7.7)

This is a pseudo jitter because at that point, the amplitude noise is not removed.

The probability that this noise is greater than 0.5 LSB is given by

$$P = \frac{1}{P} \left[1 - \operatorname{erf} \left(\frac{0.5}{\sigma_{k2}\sqrt{2}} \right) \right]$$ (7.8)

Finally, k' is determined by

$$k' = \frac{M}{4\pi} \left(\arccos \left\{ \frac{2^{N-1} - 2}{2^{N-1}} \right\} - \arccos \left\{ \frac{2^{N-1} - 1}{2^{N-1}} \right\} \right)$$
$$\times \left[1 - \operatorname{erf} \left(\frac{0.5}{\sigma_{k2}\sqrt{2}} \right) \right]$$ (7.9)

The rms value of the amplitude (σ_A) noise is calculated on $\Delta_1 - k'$ points and the obtained value is then subtracted from σ_{tot}, which leads to the rms value of the noise induced by the jitter of the two ADCs.

Calculation of the jitter

The rms value of the jitter induced noise is

$$\sigma_v = \sigma_{ji} \times \left(\frac{dv}{dt}\right)_{rms} \tag{7.10}$$

with $i = 1, 2$ or 3, depending on the ADCs considered.

To determine precisely σ_{ji}, the rms slope of the input sinewave must be calculated accurately, without taking into account the distortions introduced by the ADC.

To perform that calculation, the FFT of the acquired signal is calculated for each channel. On these two signals, only the fundamental is kept to its value, the other lines of the spectrum are set to zero. Then a reverse FFT is performed and the slope of the resulting signal is calculated for each channel. The slope calculated is the slope of the beat signal, and the slope of the input signal is then determined by

$$Slope_{in} = Slope_{beat}\frac{pM \pm 2}{2} \tag{7.11}$$

Finally, the arithmetic mean of the two slopes calculated for each channel is used to determine σ_{ji}.

2.5   Experimental results

To check the validity of the double beat method for state of the art converters, measurements were performed on a 14-bit 65 MS/s ADC from Analog Devices, the AD6644AST-65. The typical rms jitter given by the manufacturer is 0.2 ps.

All the measurements were performed on records of 65536 samples. The sampling frequency was 65 MS/s and the input frequency was set very close to 260 MHz. The value of the input frequency was calculated with $p = 3$. It was set to its highest value given the input bandwidth of the AD6644 (250 MHz typ.).

The amplitude of the input sinewaves were set very close to $FS - 0.1$ dB. For jitter measurement it is important to use a sinewave as close to Full Scale as possible, in order to maximise the effect of the jitter. For all the measurements performed to validate the double beat method, the jitters given are composed by the jitters of the two ADCs.

2.6   Sensitivity to the imbalance of the channels

For that measurement the test setup is: The two channels can be perfectly balanced by the use of the variable attenuators and of the delay lines represented

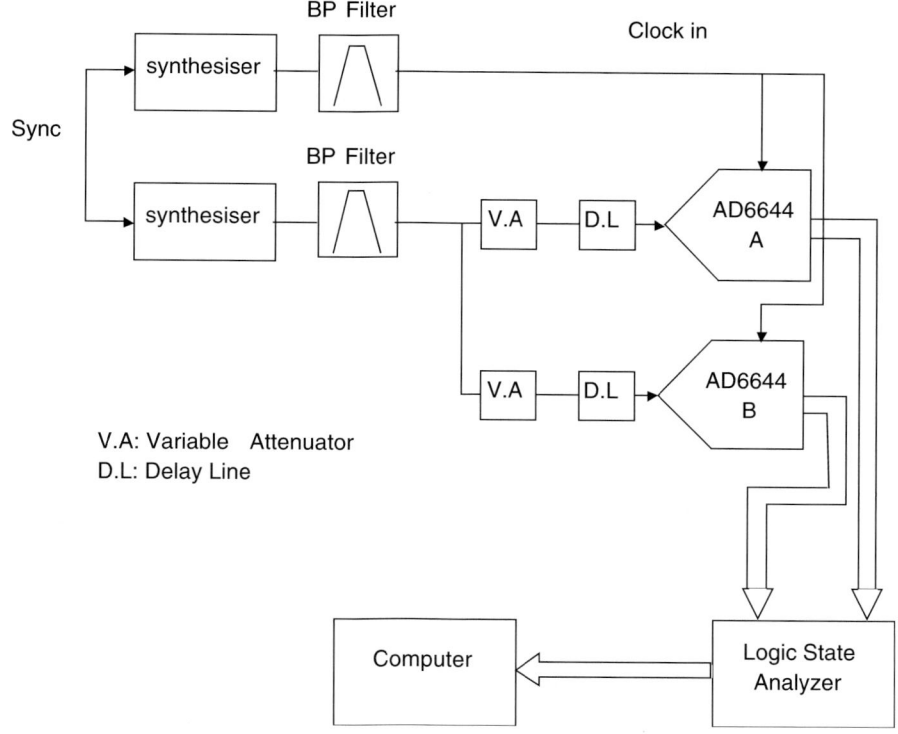

Figure 7.7. Test set up for the testing of the sensitivity to the imbalance of the channels.

in figure 7.7.

Sensitivity to phase imbalance

First, the amplitudes of the two channels were balanced and the differential phase was varied. The variation of the measured jitter as a function of the differential phase is given in the following graph: The graph above shows that the double beat method works well for a differential phase between the two channels as high as 4 degree.

Sensitivity to the amplitude imbalance

To check the sensitivity of the method to the amplitude imbalance, the differential phase was adjusted to a value lower than 1 degree. The amplitude of one channel was kept to $FS - 0.1$ dB and the amplitude of the other one was decreased from $FS - 0.1$ dB. The measured jitter is plotted as a function of the amplitude difference between the two channels: The graph above shows that the double beat method works well for an amplitude difference between the two channels as high as 0.5dB.

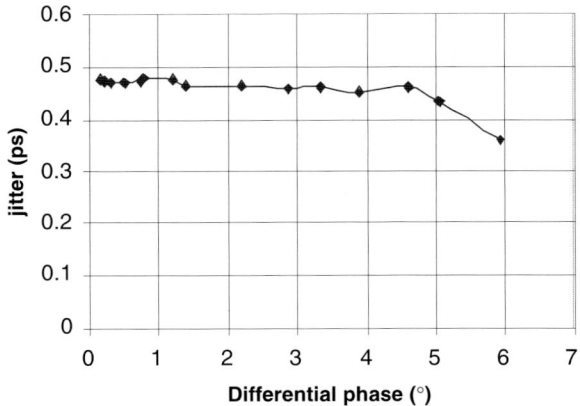

Figure 7.8.   Measured jitter as a function of the phase imbalance between the two channels.

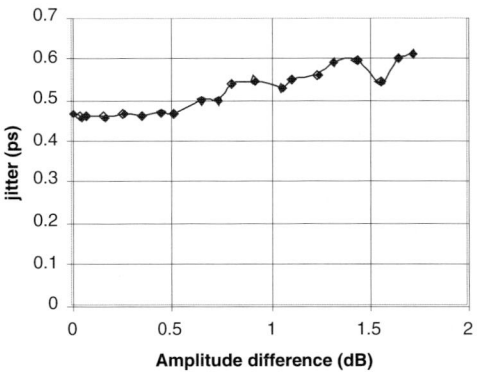

Figure 7.9.   Measured jitter as a function of the amplitude imbalance between the two channels.

These results show that to perform correct measurements with the double beat method, the phase imbalance and the amplitude imbalance between the two channels must be respectively lower than 4 degree and 0.5dB. These conditions are very easy to fulfill with classical splitters and cables.

2.7   Sensitivity to the noise of the testbench

The amplitude and phase imbalance were adjusted to values respectively lower than 0.1dB and 1degree. Additive white gaussian noise was successively coupled to the amplitude signal and to the clock signal. The noise was coupled after the band-pass filters and before the coupler used to split the signal into two channels. The jitter was measured as a function of the noise amplitude.

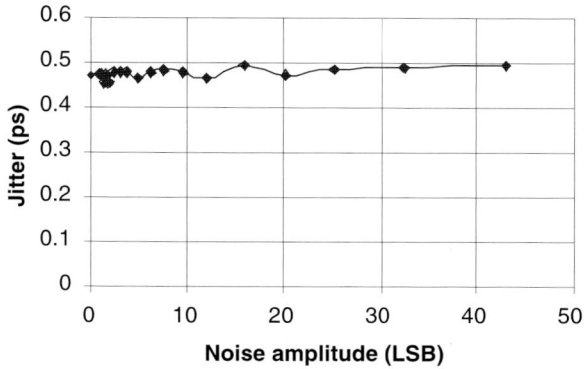

Figure 7.10. Measured jitter as a function of the additive noise on the input signal

Noise coupled to the input signal

The variation of the measured jitter is very low even for additive noise as strong as 43 LSB. A jitter increase of 4% was observed between the measurement without additive noise and the measurement with an additive noise of 43LSB.

Noise coupled to the clock signal

The variation of the measured jitter is very low even for additive noise as strong as 43 LSB. A jitter increase of 8% was noticed between the measurement without additive noise and the measurement with an additive noise of 43LSB.

The two curves above show that the subtraction technique of this method is

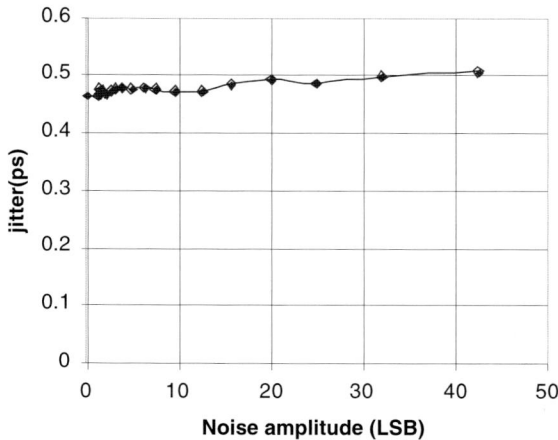

Figure 7.11. Measured jitter as a function of the additive noise on the clock signal

Table 7.1.   Jitter measurements.

	ADC A	ADC B	ADC C
Jitter (ps)	0.33	0.33	0.40

very efficient and thus that the noise of the testbench is very well cancelled by this method.

2.8   Measurement of the AD6644

The measurements were performed with three AD6644:

- two AD6644XST-65 (prototype) referred as ADCs B and C

- one AD6644AST-65 (sample) referred as ADC A.

The amplitudes of the two input signals were close to $FS - 0.1$ dB and the amplitude difference was lower than 0.1 dB. The differential phase was adjusted lower than 1 degree.

The jitters measured are listed in table 7.1

2.9   Conclusions

The measurements performed with the AD6644 ADC allowed to obtain a jitter of 0.33 ps for two components and 0.4 ps for the other one. The accuracy of the double beat method was evaluated to ± 0.3LSB in [88]. With the procedure described here an accuracy of ± 0.03 ps could be obtained.

For the sample version (AD6644AST), the SNR of the component was measured as a function of the input frequency with ultra low phase noise synthesizers. The decrease of the SNR as the input frequency increases led to a jitter of nearly 0.3 ps.

These different results show that the jitter of the AD6644AST tested is close to 0.33 ps (the typical value of the jitter of the AD6644 given by Analog Devices is 0.2 ps).

All the measurements described previously show that the double beat method is very well suitable for the measurement of very low jitters.

3.   The joint probability technique

3.1   Setup and principle

The setup to be used with this technique is depicted in figure 7.12. A unique synthesizer feeds the same signal to the analogue and encode paths of two similar ADCs. Since the input and clock signals are the same, each converter should ideally output only one code. In the real world, additional codes spread around a mean code because of the ADC internal noise and the additive noise of the testbench .

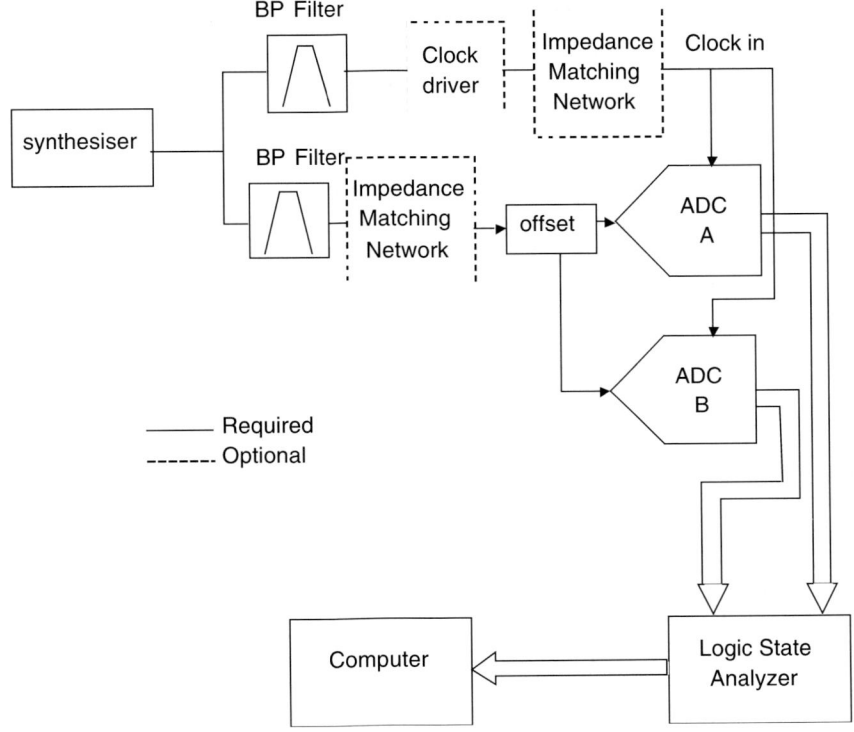

Figure 7.12. Dual channel setup for the join probability technique

Neglecting quantization noise, the noise at the output of channel A is

$$n_A = n_{ADC,A} + n_s \qquad (7.12)$$

Where n_s is the noise contribution from the test setup and $n_{ADC,A}$ is the internal noise of ADC A.

The code variance at the output of channel A is

$$\sigma_A^2 = \sigma_{ACD,a}^2 + r^2 \qquad (7.13)$$

where r^2 is the variance of the correlated noise (due to the test bench) and $\sigma_{ACD,A}^2$ is the variance of the ADC internal noise. $\sigma_{ACD,A}^2$ takes into account both the jitter induced noise and the amplitude noise of the ADC. The code variance at the output of channel B can be similarly derived.

3.2   Determination of the ADCs' internal noise

On each channel, assuming a gaussian distribution for the noise, the fraction of codes higher than a reference code k is given by (for channel A)

$$F_{nA}(\mu_A) = \frac{1}{2}\left\{1 - erf\left[\frac{T_{k+1} - \mu_A}{\sigma_A\sqrt{2}}\right]\right\} \qquad (7.14)$$

where T_{k+1} is the transition level between the codes k and $k+1$ and μ_A is the mean value of the input signal.

σ_A and σ_B are determined by least-square fitting the experimental data measured for different values of μ_A and μ_B.

The joint probability that the codes from ADCs A and B are at the same time higher than two reference codes k and j is measured. This is the joint distribution function of the noises n_A and n_B, $F_{nA,nB}$. The joint pdf is $f_{nA,nB}$.

Consider the following variables

$$n_A + n_B = 2n_s + n_{ADC,A} + n_{ADC,B} \qquad (7.15)$$

and

$$n_A - n_B = n_{ADC,A} - n_{ADC,B} \qquad (7.16)$$

Their variances are respectively

$$\sigma_{nA+nB}^2 = 4r^2 + \sigma_{ACD,A}^2 + \sigma_{ACD,B}^2 \qquad (7.17)$$

and

$$\sigma_{nA-nB}^2 = \sigma_{ACD,A}^2 + \sigma_{ACD,B}^2 \qquad (7.18)$$

The distribution function is

$$F_{nA+nB}(z) = \int\int_{\Omega_z} f_{nA,nB}(\mu_A, \mu_B)d\mu_A d\mu_B \qquad (7.19)$$

where Ω_z is the region where $\mu_A + \mu_B < z$. Similarly, the integration of $f_{nA,nB}$ over the region $\mu_A - \mu_B < z$ provides the distribution function of $n_A - n_B$. These two distribution functions can be fitted in the same way than F_{nA} and F_{nB} to determine σ_{nA+nB} and σ_{nA-nB} and therefore r. Knowing σ_A (respectively σ_B) and r, the internal noise of ADC A (respectively ADC B) can be calculated.

3.3   Extraction of the jitter of the ADC

As explained in the previous section, the internal noise of each ADC can be determined. The variance of this noise can be expressed as:

$$\sigma_{ADC,A}^2 = \sigma_{vADC,A}^2 + (2\pi f A \cos(\phi))^2 \sigma_{jADC,A}^2 \qquad (7.20)$$

where A and f are respectively the amplitude and frequency of the input sinewave and ϕ the sampling phase. $\sigma^2_{vADC,A}$ is the variance of the amplitude noise created by the ADC and $\sigma_{jADC,A}$ is the aperture uncertainty (jitter) of the ADC.

Two successive measurement of $\sigma^2_{ADC,A}$ are necessary to determine $\sigma_{jADC,A}$. The first one is performed with $\phi = \frac{\pi}{2}$ gives $\sigma^2_{vADC,A}$ and the second one is made for another value of ϕ ($\phi = 0$ maximises the effect of the jitter, it is the best value to use). Subtracting $\sigma^2_{vADC,A}$ from the second measurement and knowing A and f leads to the value of $\sigma_{jADC,A}$. Of course the same reasoning for channel B leads to the jitter of ADC B, $\sigma_{jADC,B}$.

4. Conclusion

The two methods presented in this chapter are based on a dual-channel block diagram. The use of two channels allow the additive noise created by the test-bench to be cancelled by the signal processing, which is essential to perform accurate measurements of very low jitters.

The experimental results show that the double beat technique gives very good results for the jitter measurement of state of the art high speed, high resolution ADCs. This method is also very well suitable for the measurement of very high resolution ADCs.

The joint probability technique seems to be a good solution for low-to-medium-resolution ADCs. However, this method should be evaluated experimentally for the measurement of the jitter and not only for the measurement of the ADC noise.

Chapter 8

DIFFERENTIAL GAIN AND PHASE TESTING

José Machado da Silva

Universidade do Porto, FEUP - INESC Porto
Campus da FEUP, Rua Dr Roberto Frias
4200-465 Porto, Portugal

jms@fe.up.pt

Hélio Mendonça

Universidade do Porto, FEUP - INESC Porto
Campus da FEUP, Rua Dr Roberto Frias
4200-465 Porto, Portugal

hsm@fe.up.pt

1. Introduction

Composite video encodes brightness (luminance), timing (sync), and color (chrominance) into one channel. Luminance is the voltage offset from a reference, or "black", level. Chrominance is encoded as a high-frequency (with respect to the luminance signal) sub-carrier. The average value (mid-point) of the chrominance is the luminance. The color has two "dimensions": amplitude which determines the saturation, and phase relative to a reference chrominance burst which encodes the hue. Red in NTSC[1] is shifted 103.7° from the reference, green 241.3°. Changes in amplitude or phase of the chrominance-subcarrier due to non-idealities in the sampling A/D converter directly relate to changes of color and brightness of the TV picture.

So distortion-free processing of a color signal requires that neither the amplitude nor the phase of the chrominance signal be altered as a function of the associated luminance signal.

Two parameters are commonly measured to characterize the behaviour of video ADCs - differential gain and differential phase. **Differential gain** is defined as the percentage difference between the output amplitudes of a small

---

[1] National Television System Committee

D. Dallet and J. Machado da Silva, (eds.), Dynamic Characterisation of
Analogue-to-Digital Converters, 235–242.
© 2005 *Springer. Printed in the Netherlands.*

high-frequency sinewave at two stated levels of a low-frequency signal on which it is superimposed. **Differential phase** is the difference in the output phases of a small high-frequency sinewave at two stated levels of a low-frequency signal on which it is superimposed.

2.   Test setup and hardware requirements

2.1   Test signal waveform

A test signal containing the following components has to be applied at the ADC input [7, p. 26]

- A low-frequency component at line frequency (15625Hz) which varies from blanking to 90% of the total luminance signal level (90 IRE). A staircase waveform of five to ten equal height steps is typically used.

- A high-frequency sinewave having a peak-to-peak amplitude of 40% of the luminance signal level (40 IRE). In a PAL-system (Phase Alternate Line), this sinewave shall be at a frequency of 4 433 618.75±5 Hz. In an NTSC system it shall be at a frequency of 3.58MHz.

- Normal synchronizing and blanking signals. The color burst may or may not be present.

Analogue video signals are measured in an IRE (Institute of Radio Engineers) scale. An IRE unit is defined as $\frac{1}{100}$ part of the luminance (blanking to reference white) range — an IRE unit equals 7.14 mV. Blanking level is 0 IRE units and peak white level is 100 IRE units (700 mV). IRE below blanking level is referred to as negative values — -40 IRE to +100 IRE = 1 V [7, p. 4].

The waveform of figure 8.1 shows a possible test signal.

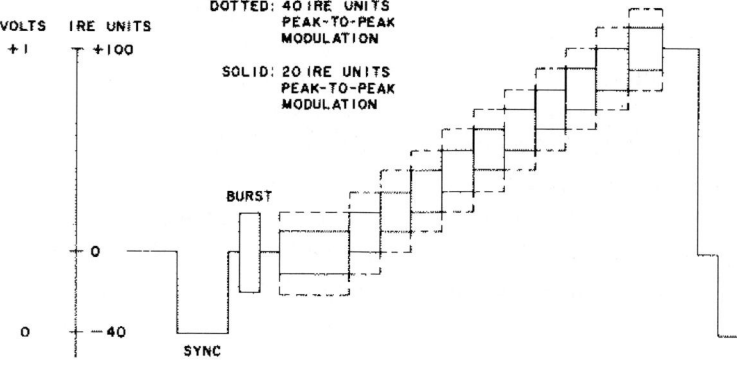

Figure 8.1.   Ten-step modulated staircase test signal.

Variations can occur in the number of staircase levels and in the amplitude of the high-frequency carrier. The synchronizing and blanking signals need not be present for ADC testing.

2.2   Test signal generation circuitry

A special circuitry is needed for driving ADC inputs. The stream of samples should not be interrupted because the phase relation of the DFT sample set would be destroyed. A circuit suitable for differential gain and phase measurement with differential inputs is shown in figure 8.2. It produces an input waveform for the ADC by superimposing a sinewave signal on a staircase as described above.

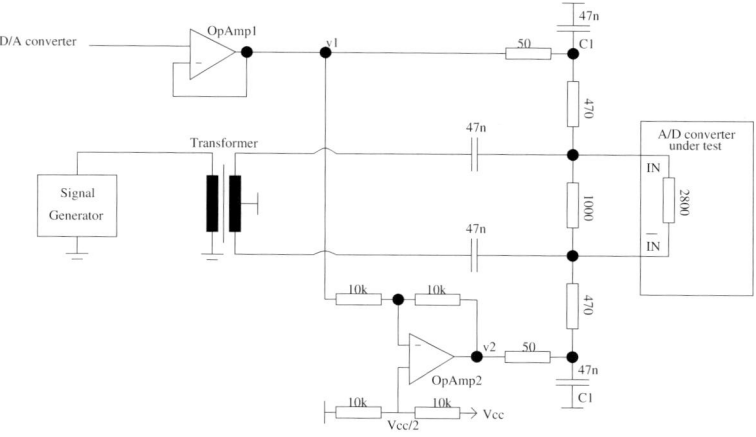

Figure 8.2.   Differential input drive circuit.

3.   Analysis

3.1   Considerations concerning digitized signals

Even in an ideal system gain and phase errors occur only due to quantization errors. Assume a standard differential gain and phase test is applied to an ideal, noise-free digital video system. Figure 8.3 shows input sinewaves (which should be thought of as superimposed on fixed voltages, according to differential gain and phase testing) each sampled three times, the samples being accurately spaced by $120°$.

Each sample is now quantized, i.e., it is represented as the nearest quantization level. It can be seen that the quantized level may differ from the true one

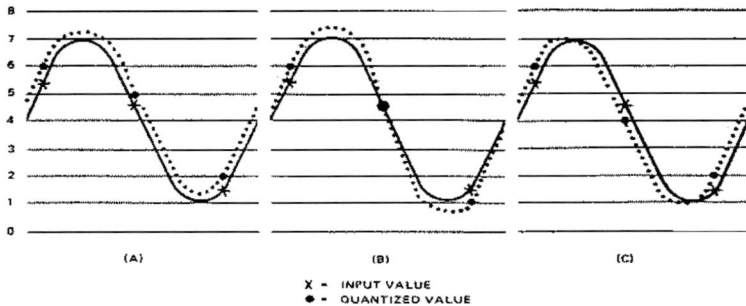

Figure 8.3.   The three cases of quantizing error.

by $\pm\frac{1}{2}$ LSB. The resulting sinewave can differ from the applied one in three ways[2]

- If all three quantized levels are equally too high (all three sample values must lie above a critical level so that its quantized value is $\frac{1}{2}$ LSB too high) the result is a DC shift to the waveform (see figure 8.3 A). In television terms: chrominance to luminance crosstalk.

- If the errors are symmetrical about the center of the sinewave, the result is a change in amplitude (see figure 8.3 B). In television terms: chrominance gain error.

- A particular form of asymmetric error can result in a phase shift without change in either amplitude or DC component (see figure 8.3 C). In television terms: chrominance phase shift.

None of these three cases is likely to occur in a practical measurement, especially when the number of samples taken is quite high. By applying oversampling, which means taking more samples per period, the chance that quantization errors accumulate to one of the three extreme cases described above, becomes even smaller. Choosing sampling and signal frequencies as noninteger multiples contributes to averaging gain and phase errors.

3.2   Measurements analysis

Figure 8.4 shows a sample test signal captured from an 11-bit ADC with differential inputs using the circuit described above. The 8 steps show some rounding at the beginning of each step. This is due to the settling time between different DC levels.

---

[2][50, p.76]

Parameters: $f_i = 4.443359375$ MHz, $f_s = 28$ MHz, N=32768, 11 bits
8 steps, 40% fullscale peak-to-peak amplitude (approximately 819 LSB)
FFT results: N_{ef}=7.804 bit, SNR=48.74dB, THD=58.98dB

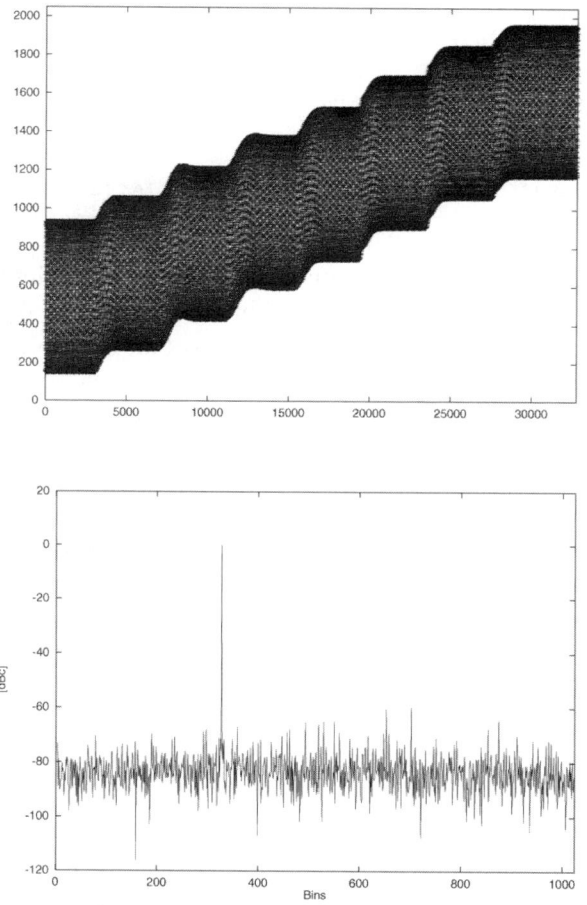

Figure 8.4. Test signal captured from the A/D converter and FFT plot of the first level.

The waveform is sampled coherently. This prohibits harmonics of the input signal from sharing the same bin in the DFT result as the fundamental due to aliasing. This could lead to amplitude and phase deviations of the fundamental and falsify the results.

For controlling the quality of the captured signal an FFT should be done over one single level of the staircase signal in order to assure that sample values have been taken correctly and there is enough spectral purity to get significant measurement results.

Table 8.1.   DFT test results.

level#	gain [LSB]	gain error [%]	phase [°]	phase error [°]
0	195.32	0	-164.06	0
1	195.50	0.092	-164.10	0.04
2	195.50	0.092	-164.14	0.08
3	195.43	0.056	-164.11	0.05
4	195.45	0.067	-164.07	0.01
5	195.43	0.056	-164.10	0.04
6	195.28	-0.020	-164.16	0.10
7	195.13	-0.097	-164.14	0.08

Simply searching minimum and maximum values in the digital output and calculating the result from these will yield inaccurate results. Applying a DFT on the digital samples has the advantage that most of the noise is neglected (since the main purpose of a DFT is the separation of spectral energy, in this case the quantization/voltage/distortion noise energy is separated from the fundamental signal energy).

According to the parameters in figure 8.4, $J=325^3$ periods are taken for DFT, with one half of the samples per DC level as input, during each DC level. The DFT is done only for bin 325 which is the bin of the fundamental frequency. The result is a complex number $X(k) = \Re e(k) + j\Im m(k)$ from which magnitude (A) and phase (φ) can be easily calculated

$$\varphi = \arctan\left(\frac{\Im m(k)}{\Re e(k)}\right) \qquad \varphi = -\pi, \cdots, +\pi \qquad (8.1)$$

$$A = \sqrt{\Re e(k)^2 + \Im m(k)^2} \qquad (8.2)$$

The DFT result of the first stage is the reference for both magnitude and phase. The other DFT results are then compared to this reference.

4.   Test results

Table 8.1 gives an overview of the DFT results from the 8 DC levels. It should be noticed that the DFT result for gain usually is not the signal amplitude in LSB. Furthermore, it should be noticed that the DFT result for phase at the first DC level is more or less random since the DFT results depend on the absolute phase relationships between clock and signal at the start of measurement.

Figure 8.6 shows graphical representations of the results listed in table 8.1.

$^3\left(\frac{f_i}{f_s}M = \frac{4.443359375}{28}2048 = 325\right)$ see (4.3) on page 86

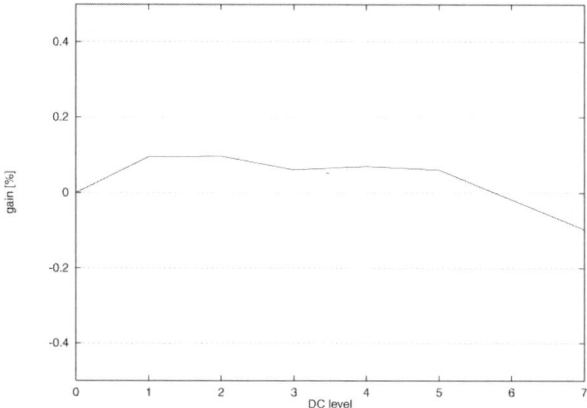

Figure 8.5. Differential gain test results.

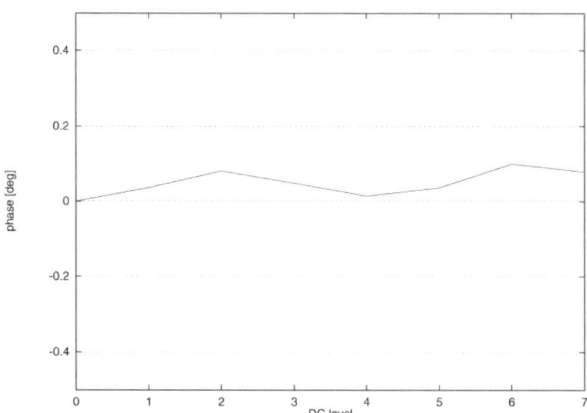

Figure 8.6. Differential phase test results.

Table 8.2. Differential gain and phase results.

calculation method	gain error [%]	phase error [°]
peak-to-peak	0.189	0.1
RMS	0.073	0.064
endpoint	0.097	0.08

5. Calculation of differential gain and phase from the test results

Table 8.2 shows the resulting gain and phase error. Simply giving the endpoint delta between the first and the last staircase level as a final specification can be misleading. It can be seen from the figure that differential gain and phase do not behave linearly with increasing DC levels (luminance levels).

Measuring the peak-to-peak error yields 0.189% for differential gain (difference of steps 2 and 7) and 0.1° for differential phase (difference of steps 0 and 6).

A specification of the RMS error yields 0.073% for differential gain ($\sqrt{\frac{0.092^2+0.092^2+0.056^2+0.067^2+0.056^2+0.020^2+0.097^2}{7}}$) and 0.064° for differential phase.

Chapter 9

STEP AND TRANSIENT RESPONSE MEASUREMENT

Giovanni Chiorboli

Dip. di Ingegneria dell'Informazione, University of Parma
Parco Area delle Scienze 181/A, 43100 Parma, Italy
giovanni.chiorboli@unipr.it

Carlo Morandi

Dip. di Ingegneria dell'Informazione, University of Parma
Parco Area delle Scienze 181/A, 43100 Parma, Italy
carlo.morandi@unipr.it

1.   Introduction

Step and transient response measurement provides some additional insights in the knowledge of the behaviour of ADCs, both in the time and frequency domains .

For instance, settling time and transition duration of step response can not be measured by the classical sinewave test. In addition, the complex (amplitude and phase) frequency response can be more quickly estimated from the Discrete Fourier Transform of the derivative of the step response rather than step-by-step varying the frequency of the input sinewave over the useful bandwidth of the ADC.

The measurement is performed by feeding the converter with a voltage step and by acquiring the response. The input step must be obviously as ideal as in the case of the measurement of a step response of an analogue linear circuit, and therefore the transition duration, the overshoot and the settling time of the input step signal have to be smaller than one-fourth of those expected from the ADC under test.

In the case of AC-coupled high-speed ADCs, the test circuit has to be designed keeping in mind that flat-frequency response and impedance matching must be guaranteed over a wider bandwidth than in the classical sinewave test.

Moreover, because of discrete-amplitude, discrete-time characteristics of ADCs, particular care should be taken for the test conditions, since aliasing, jit-

D. Dallet and J. Machado da Silva, (eds.), Dynamic Characterisation of
Analogue-to-Digital Converters, 243–254.
© 2005 *Springer. Printed in the Netherlands.*

ter, non-perfectly coherent sampling, noise and ADC nonlinearities (non linear distortion and quantization) affect the measurement accuracy.

1.1   Equivalent-time sampling

Sometimes the step response of the ADC is shorter than the minimum sampling period, and only one (or none) sample is collected in the transition, because of the small ratio of the maximum allowable sampling frequency to the ADCs' bandwidth . Moreover, small values of the ratio give rise to substantial aliasing errors in the frequency domain, since the input signal is not bandlimited. In this case it is necessary to increase the effective sampling rate by using a repetitive input signal and by choosing an appropriate value for the input repetition rate.

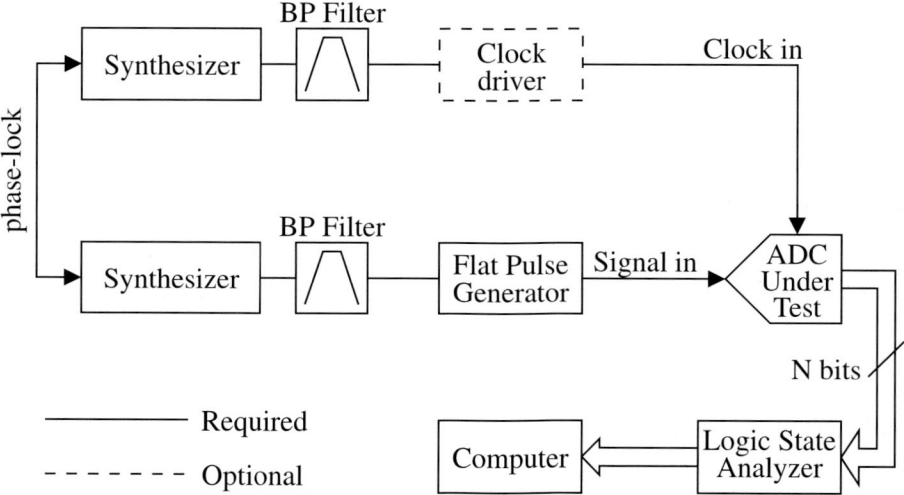

Figure 9.1.   Test setup for measuring the step response.

Figure 9.1 shows a typical test setup which can be used in the equivalent-time sampling conditions. The frequency of the second synthesizer, which determines the pulse repetition rate and drives the flat pulse generator, has to be coherent with the clock frequency, i.e.

$$r \overset{\triangle}{=} \frac{f_i}{f_s} = \frac{J}{M} \tag{9.1}$$

where M is the number of samples in a record and J is an integer number mutually prime with M. In the case of equivalent-time sampling ($J \neq 1$) the samples have to be reorganized by reshuffling with *modulo J* counting as in

(10.10). The equivalent-time becomes

$$t_{s,eq} = \frac{1}{J f_s} = \frac{1}{M f_i} \qquad (9.2)$$

and the sampling-theorem requirements can be more easily satisfied.

2. Settling time and transition duration of step response

To measure time-domain parameters the step response may be acquired in the equivalent-time domain, using a record length M and a repetition period $1/f_i$ sufficient to represent the desired parameter over the specified duration.

2.1 Settling time

Settling time t_{ST} is defined as the total time required , from the 50% point of a full-scale output transition, for the output to settle to within the specified error band ε_{ST} around the final value [6]. It is a function of the SHA's ability to track fast slewing signals or, in sigma-delta ADCs, it depends on the fact that when a full-scale step stimulates the ADC, the entire digital filter must fill with the new data before the output becomes valid. Both short-term and long-term settling times can be defined by specifying two different error bands.

The uncertainty in the settling time of the input step causes a comparable uncertainty in the corresponding estimated ADC settling time. To achieve an uncertainty lower than $1/B$, a reasonable rule-of-thumb is that the input step settles to within the given error bound ε_{ST} in a time no greater than t_{ST}/B and that, at t_{ST}, the input signal is within a bound ε/B around its final value [138].

When these requirements cannot be satisfied, it is possible, for achieving smaller uncertainties, to digitally deconvolve the output data with the known input step, as described below in section 2.3.

Digitally filtering the step response before computing settling time, for instance applying a moving average filter, is a reported technique for improving measurement repeatability [6, 138]. In effect it filters out noise and quantization errors, but it biases the estimation of settling time and should be used with great caution [138].

2.2 Transition duration of step response

The transition duration of the step response is the duration between the 10% point and the 90% point of the output transition, which can be determined by linear interpolation when only few samples are available on the transition.

Since the composite transition duration of two cascaded gaussian filters is the root-sum-of-squares of the individual transition duration

$$\tau_{r,OUT} = \sqrt{\tau_{r,1}^2 + \tau_{r,2}^2} \qquad (9.3)$$

a rule-of-thumb is that the transition duration of the input step signal is no greater than one fourth that expected of the ADC under test. However this equation must be used with some caution since it is derived for gaussian systems [38] and there is the possibility of a significant error for an ADC which has not, in general, a gaussian response.

To achieve smaller uncertainties digitally deconvolution of the output data from the known input step can be applied, as described below.

2.3 Digital deconvolution of output step response from step input signal

To correct for the non-idealities of the input step, digital deconvolution of the measured output step response from the known input step can be applied.

Assuming that the ADC under test is a linear time-invariant system followed by an ideal sampler, the measured step response $y[j]$ ($j \in [0, N-1]$) is the sampled convolution of the step input signal $x[t]$ with the impulse response $h(t)$ of the ADC. If $x[t]$ is known, it is possible to obtain the complex transfer function of the ADC, $H[n]$, as the ratio between the discrete Fourier transform (DFT) of the ADC's response, $Y[n]$, and the DFT of the input step, $X[n]$. In particular, to apply the waveform deconvolution properly, the time epoch and the pulse width must be sufficient long for essentially complete settling of the waveform at its end. The generalized extended function fast Fourier transform proposed in [39, 40] can be used for allowing spectral representations in term of impulse response. A practical alternative is the Nahman-Gans technique [55], which takes the step response out to where the settling of the waveform is ended, and turn it off mirroring the turn on data, thus making a $2 \times N$ sample sequence. DFT can be then applied after removing the average value component.

However, the direct deconvolution is an ill-posed problem, highly sensitive to measurement noise, that, in general, leads to large deviations in the reconstruction when the inverse DFT is calculated for estimating h. The solution is employing a deconvolution algorithm that low-pass filters in the frequency-domain, so as to reduce the noise-induced errors. The complex transfer function is thus estimated as

$$H[n] = \frac{Y[n]}{X[n]} R[n], \quad n = 0, \cdots, 2N - 1 \tag{9.4}$$

where $R[n]$ is the regularization filter, for instance [58]

$$R[n] = \frac{|X[n]|^2}{|X[n]|^2 + \gamma |C[n]|^2} \tag{9.5}$$

where γ is the regularization parameter and $|C[n]|^2$ is the squared magnitude of the discrete Fourier transform of the second difference operator,

$$|C[n]|^2 = 6 - 8\cos\frac{2\pi n}{2N} + 2\cos\frac{4\pi n}{2N} \qquad (9.6)$$

In practice, the value of γ can be iteratively determined to minimize a model-based approximation of the root sum of squares of the estimation error [41].

2.4   Jitter

To reduce the effect of noise, jitter and quasi-coherent sampling on the measure of the settling time and of the transition duration, several records can be acquired and averaged out. Since in each record the step transition occurs at different positions, the acquired data records must be previously *synchronized* in the time-domain by a software algorithm.

Let $y_k[\cdot]$ and $y_j[\cdot]$ be two different records which have to be synchronized before averaging. A possible solution is to estimate the correlation function $R_{kj}[i]$ between $y_k[\cdot] - \overline{y_k}$ and $y_j[\cdot] - \overline{y_j}$, where $\overline{y_k} = \sum_{n=0}^{M-1} y_k[n]$ ($\overline{y_j} = \sum_{n=0}^{M-1} y_j[n]$) represents the mean value of $y_k[\cdot]$ ($y_j[\cdot]$, respectively).

The shift between y_k and y_j can be estimated as the index τ which corresponds to the maximum of $R_{kj}[\cdot]$, $R_{kj}[\tau] = \max(R_{kj}[i] \ \forall i \in [0, M-1])$.

However, vertical signal averaging of jittered steps will give a clean-appearing, but distorted step. In particular the averaged step has slower transition than the original one. In [57] it has been shown that averaging a large number of records is equivalent to low-pass filter output data by a filter with a 3 dB bandwidth given by $0.132/\sigma_j$, where σ_j is the standard deviation of the time jitter. Under the no realistic but simpler assumption of gaussian system, averaging a large number of records in the presence of a time jitter adds an other term in the root-sum-of-squares of about $(0.35/0.132) \times \sigma_j \approx 2.65\sigma_j$. In the most cases of interest for the testing of ADCs, this contribution can be neglected. In fact, up to $20 \div 30$ ps of jitter can be tolerated for ADCs with bandwidth smaller than 1 GHz, since the term $2.65\,\sigma_j \sim 80$ ps is of minor importance ($\sim 5\%$) in a root-sum-of-squares sense. Therefore the filtering effect of jitter must be included in the estimation of the uncertainty components evaluated by other than statistical means (i.e. Type B [2] [47] but, usually, it is not necessary to deconvolve the probability density function (PDF) of the jitter from the averaged step for measuring with a sufficient accuracy the settling time and the transition duration.

If the ADC bandwidth is wider or the standard deviation of jitter is greater, so that the averaged jitter has a significant effect on the measured step response, the contribution of the jitter can be estimated and the PDF of the jitter can be deconvolved from the step response as reported in [56, 57]. In particular, since

jitter induces a voltage noise proportional to the slope of the averaged wave-
form at a certain time instant, it is possible to estimate σ_j as $\hat{\sigma}_j = \sigma_V / \tan \vartheta$,
where σ_V accounts for the estimated voltage noise in the step ramp and $\tan \vartheta$
is the slope of the measured step. In [56] it has been shown that this method
gives an asymptotic unbiased estimate for σ_j if applied to an ideal ramp when
no additive noise is present. However, if the pulse generator provides pulse-
like signals and jitter is significant with respect to the transition duration, an
asymptotic bias will result. Even if the bias can be removed by applying the
extended PDF deconvolution method in [143], a ringing effect in the time-
domain representation and a great amount of noise at high frequency in the
frequency-domain have been observed.

Finally, after removing the mean value of the jitter-induced noise and the
bias in the estimation of the transition duration, the jitter-induced noise can be
modeled as a non-stationary zero mean additive white noise, which affects the
measurement repeatability [42]. The uncertainty contributions of the random
component of the averaged jitter can be determined by statistical means by
measuring the variance of the noise at a certain time instance and then, by ex-
trapolating the variance of the non-stationary jitter related noise for the whole
record [47]. Finally the jitter-induced noise variance can be propagated to the
output of the inverse filter in order to obtain an estimation of the uncertainty
[42].

2.5   Quasi-coherent sampling

As observed in Section 5.2.4, perfect coherence is never met because of
the finite frequency resolution of the synthesizers, and an error is done in the
frequency ratio, $\Delta r = f_i / f_s - J/M$, which yields a equivalent-time error
with a maximum value equal to

$$\Delta t_{s,eq} = MJ\, t_{s,eq} \frac{\Delta r}{r} = \frac{M}{f_s} \frac{\Delta r}{r} \quad \text{if} \quad \frac{\Delta r}{r} < \frac{1}{J(M-1)} \quad (9.7)$$

Assume for instance that f_s is exactly known and that P points have to be
acquired in the transition τ_r; an equivalent sampling frequency $f_{s,eq} = P/\tau_r$
is needed. The maximum sampling time error becomes

$$\Delta t_{s,eq} = \frac{M^2}{P} \tau_r \frac{\Delta f_i}{f_s} \quad (9.8)$$

where Δf_i is the frequency resolution of the second synthesizer. It is apparent
that the sampling time accuracy decreases as the number of acquired samples increases. However, when M is lowered, frequency resolution, given by
$1/(t_{s,eq}M)$, worsen.

Since the transition duration estimates are particularly influenced by timebase
errors [121], it is good practice to correct the data as much as possible before

applying deconvolution, which can amplify even small errors. This can be obtained by a calibration procedure in the time-domain. Let disconnect the flat-pulse generator and directly feed the ADC with the sinewave provided by the second synthesizer. A good approximation of the exact frequency ratio $\hat{r}$ can be estimated by the classical time-domain sinewave technique. Once the pulse generator is re-inserted in the step-response measurement setup, the acquired data can be reshuffled by a *modulo* $(\hat{r}\,M)$ counting process. If it is not possible to obtain an accurate estimate $\hat{r}$, a computer simulation of a single pole response with cutoff frequency equal to the nominal value of the ADC and a frequency ratio error Δr, can provide some guidelines for estimating the frequency response errors due to Δr [138].

3. Frequency response measurement

From the measurement of the step response it is possible to obtain some insight into the frequency behaviour of the ADC, as the analog bandwidth and the complex frequency response . The complex impulse response is necessary for instance for evaluating linear distortion or for a correct calibration of the ADC, since it allows knowledge of gain and phase errors versus frequency.

In [137] it has been demonstrated that the frequency response can be effectively estimated from discrete-time, discrete-amplitude step response measurements by numerical differentiation of the step response. Both the algorithm and the error sources has been deeply analyzed in literature [137, 138, 46, 24].

In figure 9.2 it is shown the flow diagram of the discrete frequency response estimation algorithm [69]. A fast step input signal $x(t)$ is applied to the input of the ADC and the step response $y(t)$ is acquired in a record at the output at the sampling time kT_s. It is preferable to average more records for reducing the random contribution of noise, jitter and quantization error.

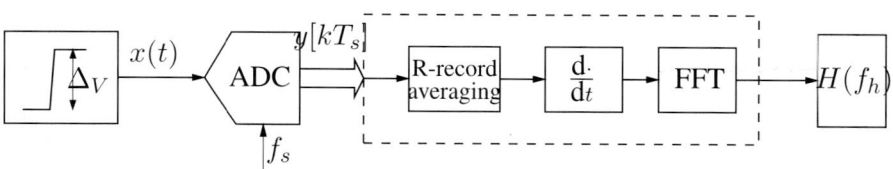

Figure 9.2. Flow diagram of the discrete frequency response estimation algorithm.

The complex frequency response $H(f)$ is the continuous-time Fourier transform of the impulse response $h(t)$. $H(f)$ can be estimated by applying the discrete Fourier transform to the time derivative of the step response, $h(t)$, which can be obtained by convolving the step response with a differentiation filter or by using the first difference operator. Since in the first case, if the length of the differentiation filter is M_d, the M samples of the step response produce an impulse response with only $(M - M_d)$ samples, it should be preferable using the simpler first difference operator, which yields a M-sample impulse response.

As stated before, care must be taken over the length of the record for essentially complete settling of the waveform at its end. Constant padding (zero padding) can be applied before (after, respectively) the derivative of the step response, for enhancing the spectral resolution of the DFT [40]. Moreover, when M is not an even power of two, padding allows a more efficient Fast Fourier Transform, without affecting the accuracy of the frequency response estimate.

The first difference of the samples provides an estimate of the impulse response given by [137]

$$h[k] \triangleq h(kT_s) = \frac{1}{\Delta_V} \frac{\mathrm{d}(y(kT_s))}{\mathrm{d}t}$$

$$\approx \begin{cases} \left(\frac{y[k+1]-y[k]}{\Delta_V T_s} \right) & \text{for } k = 0, 1, \cdots, M-2, \\ \left(\frac{y[k]-y[k-1]}{\Delta_V T_s} \right) & \text{for } k = M-1, \end{cases} \quad (9.9)$$

where Δ_V is the amplitude of the voltage step.

The DFT of the impulse response has to be calculated using the rectangular window and must be multiplied by the value of the sampling period T_s to provide an estimate of $H(f)$ at discrete frequencies $f_h = h/(MT_s)$

$$\hat{H}(f_h) = T_s \sum_{k=0}^{M-1} h[k] \exp\left(-j2\pi h \frac{k}{M} \right), \quad h = 0, 1, \cdots, \frac{M}{2} \quad (9.10)$$

Notice that the phase spectrum provided by DFT is typically wrapped, since only the remainder after dividing by 2π is given. This partly depends on the position of the step transition in the original record $y[k]$, that is an arbitrary quantity. Since only the portion of the phase spectrum that is not linearly related to frequency gives some information about the ADC, the nonlinear phase portion of the phase response can be highlighted by unwrapping the phase, i.e. by subtracting 2π after each 2π discontinuity [6].

As reported above, the error sources affecting the frequency response estimates are aliasing, noise, jitter, quasi-coherent sampling, discrete derivative and the non-ideal input step [138, 69, 24, 46].

3.1   Aliasing and first differencing

Because of the finite sampling density and the wide band occupation of the input signal, aliasing is always present in the discrete frequency response measurement, since the power in the input signal at frequencies above the Nyquist frequency folds back and adds to the power at lower frequencies. Moreover, the discrete derivative of the step response provides only an approximation of the impulse response, since differentiation corresponds to multiplying by $j2\pi f$ in the frequency domain, while the first difference operation corresponds to multiplying by $2j\sin(\pi f T_s)/T_s$. It is apparent that the error decreases as the sampling frequency increases. It is therefore desirable to increase the sampling rate by the equivalent-time sampling method previously described.

Since the power that folds back depends on the roll-off of the frequency response at frequencies above the Nyquist frequency, it is necessary to make some hypothesis about the frequency response of the ADC for quantifying the contribution of aliasing to the systematic uncertainty. A more conservative assumption is that the ADC has a single pole frequency response, so that the frequency response rolls-off at a rate proportional to f^{-1} for large f. In this case the error in the frequency response is given by [137]

$$|\hat{H}(f) - H(f)| \approx 4\,(BW\,t_{s,eq})\,(f\,t_{s,eq}) \qquad (9.11)$$

where BW is the 3 dB bandwidth of the system in units inverse of the equivalent-time $t_{s,eq}$. In many cases this estimate is excessively conservative, since the system has a higher order response. If, for instance, a more accurate model is given by a two-pole system with equal time constants, the error in the frequency response becomes [24]

$$|\hat{H}(f) - H(f)| \approx 8.4\,(BW\,t_{s,eq})^2\,(f\,t_{s,eq}) \qquad (9.12)$$

Figure 9.3 shows these two error bounds for $BW = 128$ and $M = 2048$.

3.2 Additive noise

Quantization and additive noise, which can be modeled as a white noise added to the step response, are amplified and high-pass filtered by the discrete derivative operation. If additive noise is smaller than quantization noise and the system has a single-pole response, the error contribution is given by [137]

$$|\hat{H}(f) - H(f)| \approx \frac{1}{\Delta_V}\sqrt{\pi\frac{M}{2}}\,\sin\left(\pi f\,t_{s,eq}\right) \qquad (9.13)$$

In the case of a two-pole system with equal time constants and additive noise with standard deviation σ_n greater than the quantization noise, the error becomes [24]

$$|\hat{H}(f) - H(f)| \approx \frac{\sigma_n}{\Delta_V}\sqrt{2 + (2\sin\left(\pi f\,t_{s,eq}\right))^2\,(M-2)} \qquad (9.14)$$

It is apparent from the above equations that the measurement uncertainty increases for small input step signals. Figure 9.4 shows the effect of quantization

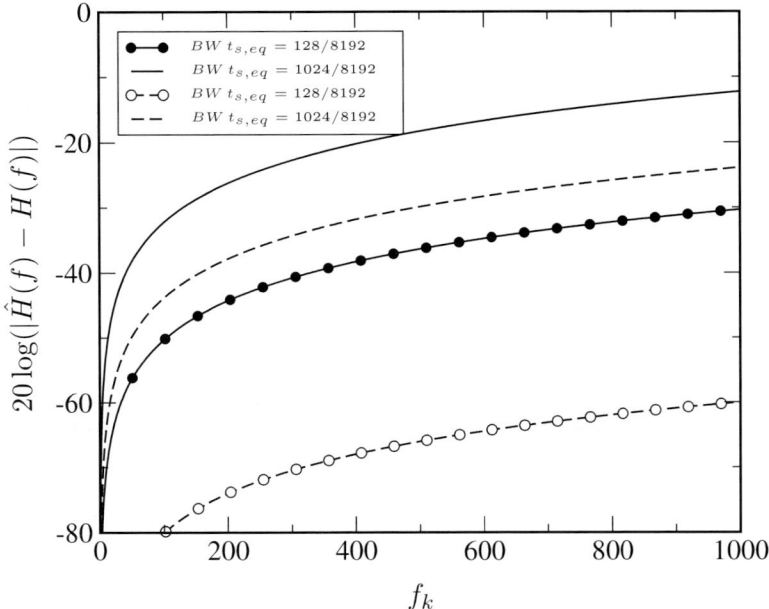

Figure 9.3. The predicted error due to aliasing in the case of single-pole (solid line) or two-pole response (dashed line). A step response with 2048 samples was assumed.

noise (additive noise is negligible with respect quantization noise) for some different values of the input step; for instance, the estimation of the bandwidth of the ADC is questionable for small value of the step and, due to quantization and additive noise, large step signals seems to be preferable. However, care must be taken that nonlinear distortion does not affect the measurement.

Averaging R records before the discrete derivation and the DFT reduces the error by a factor $\sqrt{R}$.

3.3   Jitter

As reported in the previous section, time jitter leads to a systematic error that is not removed by averaging many step response measurements. A frequency response calculated after averaging a very large number of step responses differs from the true one since it is multiplied by the Fourier transform of the PDF of the jitter. This error can be partially corrected if the jitter is measured. However, only a finite number of records are averaged, and a random uncertainty contribution is still present. It has been estimated [24] that this contribution is, for a two-pole system with equal time constants,

$$|\hat{H}(f) - H(f)| \approx \frac{9.8}{\sqrt{R}} \sqrt{(BW\, t_{s,eq})} \; \frac{\sigma_j}{\pi t_{s,eq}} \; \sin(\pi f t_{s,eq}) \qquad (9.15)$$

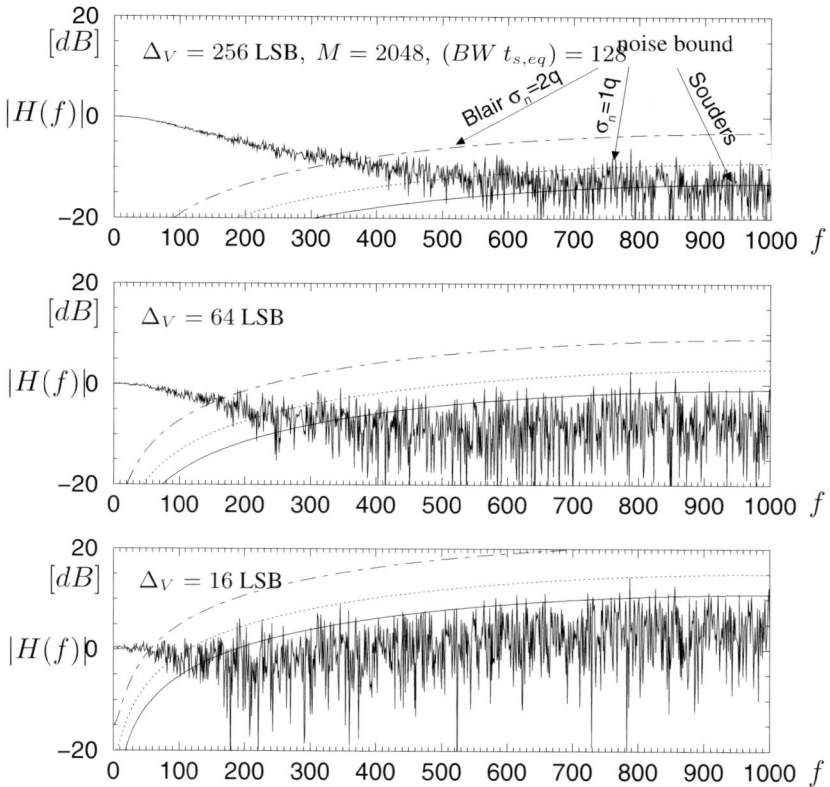

Figure 9.4. The predicted error due to additive noise and quantization in the case of a single-pole response.

where R is the number of step response averaged. Notice that σ_j must account also for the error due to the quasi-coherent sampling, which is uniformly distributed between $-\Delta t_{s,eq}$ and $+\Delta t_{s,eq}$. If the systematic contribution of jitter is not corrected, it propagates to the frequency response as described below.

3.4 Effect of the uncertainty in the knowledge of the input step

The input step signal is usually known with finite uncertainty $\pm u(t)$. The systematic uncertainty in the knowledge of the input step signal causes an uncertainty in the frequency response estimation at the discrete frequency f_h which is usually lower than [46]

$$\left| \frac{2\pi h}{M} U[h] \right| \quad \text{for } h = 0, 1, \cdots, \frac{M}{2} \tag{9.16}$$

where $U[h]$ is the DFT of the uncertainty $u[h] \triangleq u(ht_{s,eq})$.

A more conservative bound reported in the same paper is obtained by the application of the Parseval's Theorem and is given by

$$\left| \frac{2\pi h}{M} \right| \sqrt{\frac{M u_{\Sigma 2}}{2}} \quad \text{for } h = 1, 2, \cdots, \left(\frac{M}{2} - 1 \right) \tag{9.17}$$

and

$$\left| \frac{2\pi h}{M} \right| \sqrt{M u_{\Sigma 2}} \quad \text{for } h = 0, \frac{M}{2} \tag{9.18}$$

where $u_{\Sigma 2} \triangleq \sum_{k=0}^{M-1} (u[k])^2$ is the summed energy of $u[k]$.

Chapter 10

HYSTERESIS MEASUREMENT

Giovanni Chiorboli

Dip. di Ingegneria dell'Informazione, University of Parma
Parco Area delle Scienze 181/A, 43100 Parma, Italy
giovanni.chiorboli@unipr.it

Carlo Morandi

Dip. di Ingegneria dell'Informazione, University of Parma
Parco Area delle Scienze 181/A, 43100 Parma, Italy
carlo.morandi@unipr.it

1.   Introduction

Hysteresis accounts for the different behaviour of the converter when the input signal has a positive or, respectively, negative slope (history dependence or memory effect) . It is defined as the difference between the values that the transition levels assume when they are approached from the rights side rather than from the left side.

The histogram test method, which is classically used for estimating the conversion characteristic, can be easily adapted for measuring also the hysteresis. To this purpose, it is necessary to collect two different cumulative histograms as described below; the first, $H_{C_\uparrow}$, accounts for the output codes which are acquired in the positive slope, and the second, $H_{C_\downarrow}$, for those in the negative one [106].

From the two histograms, two different conversion characteristics are obtained, $T_\uparrow$ and $T_\downarrow$, and the hysteresis of the ith transition level is finally estimated as

$$HYS[i] = \frac{G(T_\uparrow[i] - T_\downarrow[i])}{Q}$$

where G is the gain error, estimated as explained in section 5.9 and Q is the nominal code bin width.

D. Dallet and J. Machado da Silva, (eds.), *Dynamic Characterisation of*
Analogue-to-Digital Converters, 255–264.
© 2005 *Springer. Printed in the Netherlands.*

2.   Test conditions

As in the classical histogram test, a sinewave

$$v(t) = A\cos(2\pi f_i t + \varphi) + C \tag{10.1}$$

is applied to the input of the N-bit ADC, R records of M samples each are collected and the code distribution is used for estimating the conversion characteristic.

The algorithms used for building the two histograms will be described in the next section. Once the two histograms have been collected, the $2^N - 1$ transition levels are given by

$$T_\uparrow[i] = C - A\cos\left(2\pi \frac{H_{C_\uparrow}[i-1]}{S}\right) \tag{10.2}$$

and by

$$T_\downarrow[i] = C - A\cos\left(2\pi \frac{H_{C_\downarrow}[i-1]}{S}\right) \tag{10.3}$$

where $S = RM$ is the total number of samples. Notice the factor 2 which multiplies π in these equations, differently from (5.121), where the transition level $T[i]$ was estimated.

The amplitude A, the input-to-clock frequency ratio, f_i/f_s, the number M of samples in a record and the number of records R have to be chosen so as to minimize the contribution of systematic and random errors.

Systematic errors arise from distortion and additive noise which should be reduced as more as possible by filtering, as in the classical sinewave tests. However, their residual effects can be compensated by slightly overdriving the ADC. Since in the hysteresis measurement two close transition levels are concerned, the effect of distortion and noise on hysteresis measurement is similar to that they have on DNL measurement.

Therefore, in order to ensure that the maximum systematic error, expressed in non-dimensional form as a fraction of the nominal code bin width Q, does not exceed E_{pdf}, the overdrive $V_{OD} = V_{OD+} = V_{OD-}$ shall be chosen so that

$$V_{OD} \geq \max\left\{3\sigma_{add}, \ \sigma_{add}\sqrt{1.43\frac{3}{8E_{pdf}}}\right\} \tag{10.4}$$

where σ_{add} is the rms value of the input referred additive B noise.

Moreover, if E_{dist} is the maximum admitted systematic error contribution of the harmonic distortion expressed in nominal code bin widths Q, then it is required that

$$\sqrt{\frac{2}{QA}} \left(\sqrt{1 + \frac{V_{OD}}{Q}} - \sqrt{\frac{V_{OD}}{Q}} \right) \sum_{i=2}^{h} i A_i \leq E_{dist} \qquad (10.5)$$

where A_i are the amplitudes of the h most relevant harmonics.

In the practical cases the most severe requirement is that of (10.4).

Once the systematic effects have been accounted for, the input and the sampling frequencies (f_i and f_s, respectively), the number of samples M in a record and the number R of records must be selected, according to the section 5.5.

In particular, in order to guarantees satisfactory coherent sampling, f_i/f_s has to be chosen so that

$$\frac{f_i}{f_s} = \frac{J}{M} \qquad (10.6)$$

with J that should be an integer number mutually prime with M. Because of the finite frequency resolution of the two synthesizers, it is not possible to exactly satisfy this requirement. However, it is sufficient that the ratio $r = f_i/f_s$ does not deviate from the above nominal value by more than

$$\frac{|\Delta r|}{r} \leq \frac{1}{2JM}. \qquad (10.7)$$

This limit guarantees that the contribution to the variance in the number of counts of the cumulated code histograms is smaller than 0.25, so that the below equation (10.8) applies.

The number of records R must be finally chosen so that

$$R \geq 2 \left[\frac{2^{N-1} k_u}{B} \right]^2 \left[\frac{c \, 2\pi}{M} \right] \left\{ 1.13 \left[\frac{\sigma^*}{V_{rir}} + \frac{c}{2} \sigma_\phi \right] + 0.25 \left[\frac{c \, 2\pi}{M} \right] \right\} \qquad (10.8)$$

where the symbols are the same of (5.101). In this equation, the factor 2 which multiplies $c\pi$ is due to the fact that only $M/2$ samples per record are approximatively collected in the histograms $H_{C\uparrow}$ and $H_{C\downarrow}$.

3. A practical case

Consider for instance a 12-bit ADC, with $f_s = 100$ MHz and $f_i \approx 50$ MHz. Let $\sigma_{add} = 1.1$ LSB and $\sigma_\phi = 2 \cdot 10^{-4}$ represent the total additive noise and, respectively, the total phase noise of the system. Let suppose that the clock signal is provided by a fixed oscillator while the input signal is synthesized by a synthesizer with 0.01 Hz of frequency resolution, so that

$$\frac{\Delta r}{r} = \frac{\Delta f_i}{f_i} = \frac{0.01}{50 \; 10^6} = 2 \; 10^{-10} \qquad (10.9)$$

From (10.7) it must be verified that $JM \leq 2.5 \ 10^9$. A possible choice is $M = 65536$ and $J = 32001$ ($f_i = 48829650.88$ Hz).

An overdrive of approximatively 5 LSB is sufficient to make negligible the contribution of the systematic errors, as demonstrated by (10.4) and (10.5).

Finally, from (10.8), $R = 30$ records are sufficient for estimating the maximum value of the hysteresis with an uncertainty approximatively of 0.2 LSB and a confidence level of 99% (from table 5.2 $k_u = Z_{0,.99} = 2.58$).

4.   Collection of samples in $H_{C\uparrow}$ and $H_{C\downarrow}$

The most intriguing problem in the measurement of hysteresis is how to separate the two histograms, since a relevant contribution to the measurement uncertainty may be due to the possibility of classifying a sample in the histogram $H_{C\uparrow}$ rather than in $H_{C\downarrow}$ and vice-versa. Therefore, the algorithm should be as more as possible insensitive to additive noise and jitter and to finite frequency resolution, and should be usable above the Nyquist frequency so as at low input frequency.

In practice, it is convenient to take advantage of the characteristic of the test setup, since the input signal synthesizer and the clock synthesizer are, in general, phase-locked. This fact allows reorganizing the acquired samples in a single period of the equivalent sampling phase. Reorganizing samples is in general useful since, in some cases, it is very difficult to estimate the signal slope by differentiating the rude data so as they are acquired. For instance in figure 10.1 it is shown the case of f_i close to the Nyquist limit, and it is apparent that it is not possible to argue the slope from the sign of $(code[k+1] - code[k])$, since the next sample $code[k + 1]$ belongs to the different slope.

In coherent sampling the acquired data can be reorganized by reshuffling with a *modulo J* counting. Once the M samples are shuffled by the *modulo J* counting, the new address of the kth sample in the vector becomes

$$k = k \ J - M \ \text{floor} \left(k \ \frac{J}{M} \right) \tag{10.10}$$

Figure 10.2 shows the resulting data in the particular case of $M = 65536$, $J = 32001$.

In order to correctly reshuffle the data, it is necessary to verify (10.6) with a sufficient accuracy. If it is not possible, the reorganized data might become unusable, as shown in figure 10.3, where a value $\Delta r/r = 0.01/32001$, approximately 1310 times the limit reported in (10.7), is used.

However, when (10.7) is verified, the reorganization of the data in a single period by the *modulo J* counting is always correctly performed.

Once the M data in the record are reorganized in a single period, it should be easy, in the ideal case of no noise and jitter, determining the conventional position of the maximum (minimum) of the signal, for instance as the median

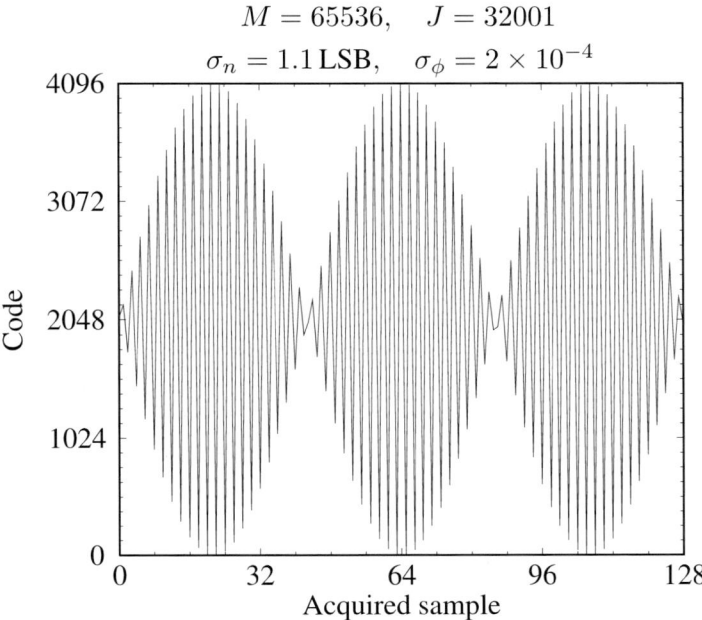

$$M = 65536, \quad J = 32001$$
$$\sigma_n = 1.1\,\text{LSB}, \quad \sigma_\phi = 2 \times 10^{-4}$$

Figure 10.1. For high-frequency input signals it is not possible to directly determine to which slope a sample belongs.

index between the indexes of the highest codes (lowest codes, respectively). However, since data are always corrupted by noise and jitter, as shown for instance in figure 10.4, it seems better to estimate the index of the maximum (minimum) as the centroid of the highest codes (lowest codes, respectively).

An example of pseudo-code for estimating the index of the maximum is reported below in pseudo-code.

```
max = 0
count_max=0
FOR k=1:M
        IF code[k] = code_max
            max = max + k
            count_max = count_max + 1
        ENDIF
ENDFOR

max = max / count_max
```

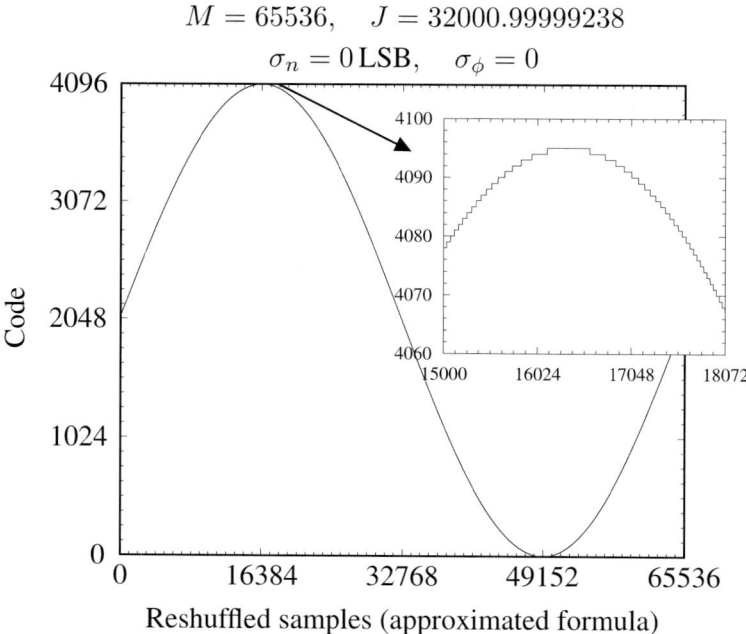

Figure 10.2. Once reshuffled according to (10.10), with $k = k\,32001 - 65536\,\mathrm{floor}(k\,32001/65536)$, the samples can be easily assigned to $H_{C\uparrow}$ and $H_{C\downarrow}$.

Finally, the estimates of the indexes of the maximum and of the minimum allow correctly classifying codes in $H_{C\uparrow}$ or in $H_{C\downarrow}$. In fact, all the samples $code[k]$ with an index k higher than the index of the minimum and lower than the index of the maximum belong to the positive slope and must be classified in $H_{C\uparrow}$, and the others must be classified in $H_{C\downarrow}$.

5.   Some warning

Notice that the hysteresis has been defined as the difference between the two output values which, on average, correspond to the same value assumed by the input signal with positive and, respectively, negative slope. This definition can be used without caution only in the ideal case of zero delay between the input V_{IN} and the output V_{OUT}. In fact, in the presence of a constant delay τ_0, hysteresis should be estimated from the input-output characteristic $V_{OUT}(t) = F(V_{IN}(t - \tau_0))$ rather than from $V_{OUT}(t) = F(V_{IN}(t))$. However, in the dynamic test of ADCs, the input signal is not known and the delay t_0 cannot be measured. The up and down branches of the conversion characteristic provided by the above modified histogram test, correspond to the transfer characteristic

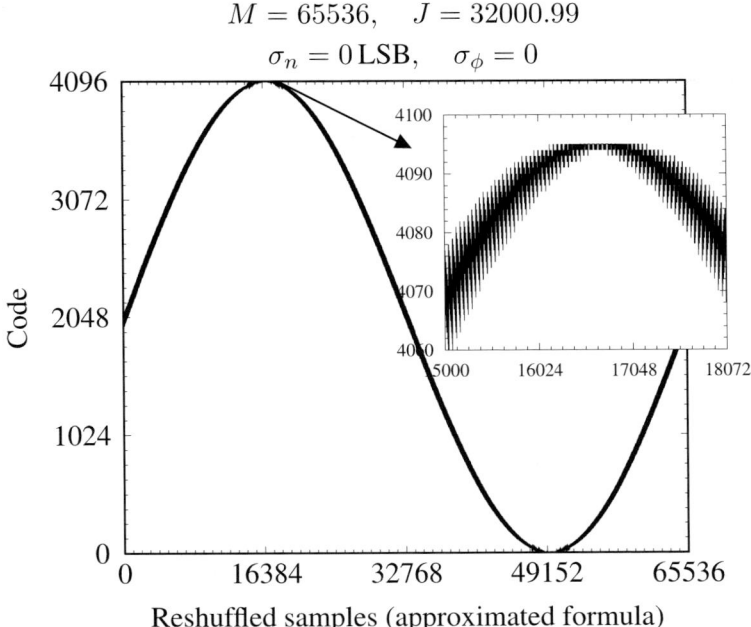

Figure 10.3. When a not integer number J of periods are collected in a record of M samples, reshuffling can be unusable.

between an ideal pure sinewave and the distorted output with the maximum (minimum) in phase with the maximum (minimum, respectively) of the input sinusoid. Therefore, in the presence of distortion, these two branches can result in an overall elliptic shape, as shown in the part b) of figure 10.5, which reports the measured deviation from the ideal conversion characteristic, for a 14 bit ADC stimulated by an input sinewave at a frequency very close to the bandwidth limit of the converter.

In order to better understand this effect, consider that the response of the converter to an input sinewave is a combination of several harmonic terms

$$V_{OUT}(t_k) = A_0 + \sum_{h=1}^{h_{\max}} A_h \sin\left(2\pi \frac{f_i}{f_s} k + \psi_h\right) \qquad (10.11)$$

Hysteresis could be estimated in the time domain as proposed in [14], by fitting the output data with (10.11) and by retaining as an estimate $\hat{V}_{IN}(t_k)$ only the fundamental tone plus the DC, $A_0 + A_1 \sin\left(2\pi f_i/f_s k + \psi_1\right)$. The up and down transition levels ($T_\uparrow[i]$ and $T_\downarrow[i]$, respectively) are then estimated as the two values that $\hat{V}_{IN}$ assumes when $V_{OUT} = T_{nom}[i]$, where $T_{nom}[i]$ is the

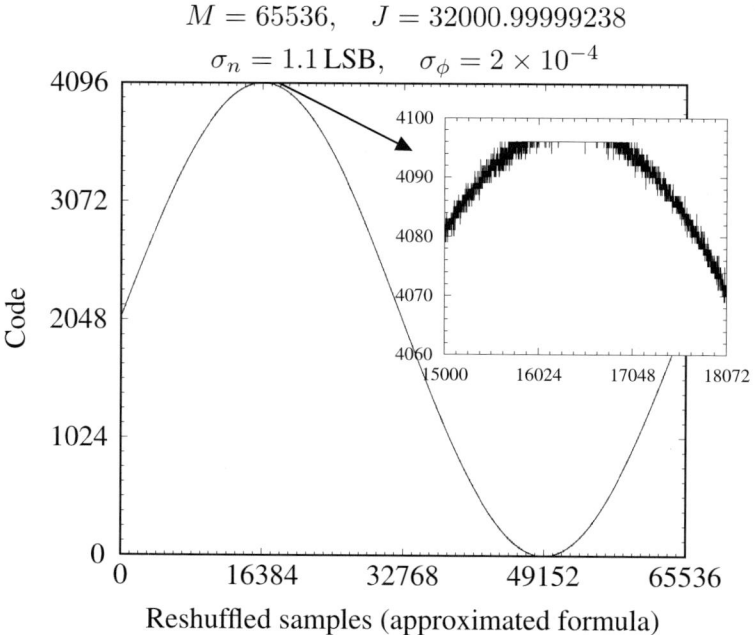

Reshuffled samples (approximated formula)

Figure 10.4.   When noise and jitter are present, the correct position of the maximum of the signal can be estimated by calculating the centroid of the highest codes.

nominal value of the ith transition level.

This measurement algorithm provides a quite different estimate of the hysteresis, as shown in the part a) of figure 10.5, where the solid line represents the output error $V_{OUT}(t_k) - \hat{V}_{IN}(t_k)$ versus the estimated input $\hat{V}_{IN}(t_k)$. Since in the time-domain the delay τ between the maxima of $\hat{V}_{IN}$ and V_{OUT} can be easily estimated, it is possible to draw also $V_{OUT}(t_k) - \hat{V}_{IN}(t_k - \tau)$, obtaining a characteristic (dash-dotted line) with an elliptic shape similar to the one reported in the part b) of the figure.

This leads into a subtle discussion on definition of hysteresis. If, for a particular architecture, it is possible to guarantee that this delay has a constant value for all the admitted values of the input frequency and amplitude, the time-domain approach seems to be preferable, since it is unaffected by this delay contribution. Nevertheless, in particular for high-speed converters, it is known that some nonlinear effects can contribute an amplitude dependent delay in addition to the harmonic distortion. For instance, the simultaneous frequency and amplitude limitation operated by latch-comparators in flash converters yields a

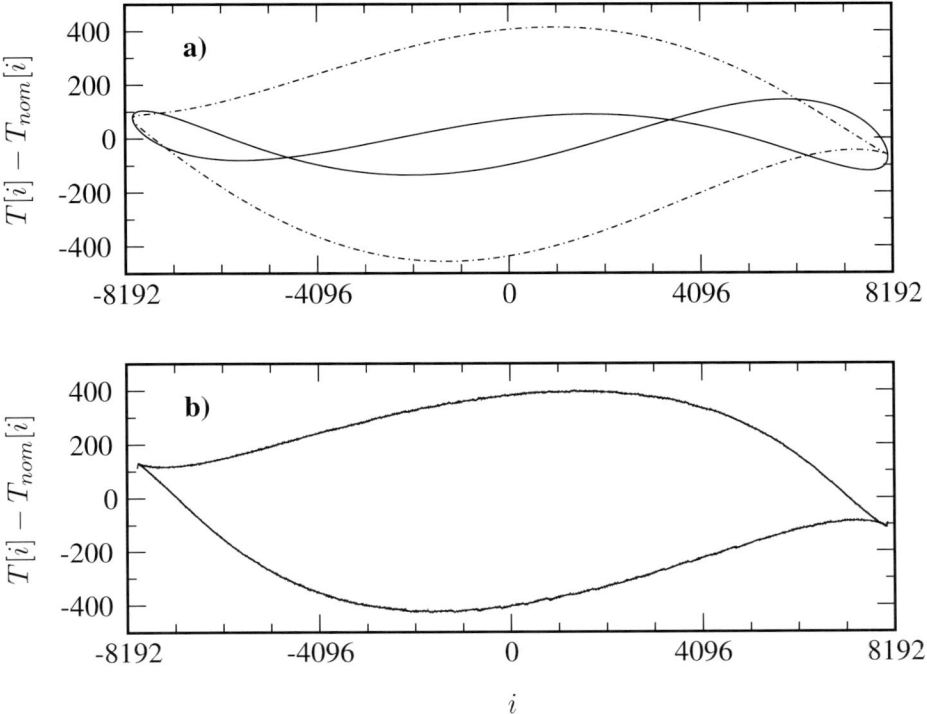

Figure 10.5. The estimated error in the two branches of the conversion characteristic. a): time-domain representation of the harmonic distortion terms as a function of the fundamental tone (solid line) and of a sinewave with the same characteristics, but slightly delayed (dash-dotted line). b): deviation from the nominal conversion characteristic estimated by the modified histogram test.

delay varying with the signal slope, so that the distorted output, apart from the quantization and the random effects, is given by [142]

$$V_{OUT} = C + A \left[\sin(2\pi f_i t + \varphi) + \frac{8}{3\pi} \delta t_d 2\pi f_i \cos(2\pi f_i t + \varphi) \right.$$
$$\left. + \frac{2}{3\pi} \delta t_d 2\pi f_i \cos(6\pi f_i t + \varphi) \right] \qquad (10.12)$$

where δt_d is the maximum delay variation. In this case, the quadrature term represents a deviation from the ideality, since δt_d nonlinearly depends on the signal amplitude. However, the time-domain fitting approach proposed in [14] will neglect this non-linear contribution.

Therefore, when there is the possibility that the delay depends on the amplitude and frequency of the signal, it is preferable to use the described statistical technique, which is sensitive to the quadrature term also, and to analyse the dependence of the measured hysteresis upon frequency.

References

[1] J. J. Corcoran, T. Hornak, and P. B. Skov, "A High-Resolution Error Plotter for Analog-to-Digital Converters", IEEE Transactions on Instrumentation and Measurement, vol. 24, no. 4, December 1975.

[2] ISO, "Guide to the Expression of Uncertainty in Measurement", ISO Technical Advisory Group 4 (TAG 4), Working Group 3 (WG 3), Geneve, Switzerland, 1995.

[3] *IEC International Standard 60748-4 – Semiconductor Devices - Integrated Circuirs*, CEI/IEC 60748-4, second edition, 1997.

[4] *IEEE Standard Dictionary of Electrical and Electronics Terms*, 3rd edition, IEEE, New York, 1984.

[5] *IEEE Standard for Digitizing Waveform Recorders*, IEEE Std 1057-1994, December 1994.

[6] *IEEE-Std-1241 - Standard for Terminology and Test Methods for Analog-to-Digital Converters*, June 2001. IEEE

[7] Electronic Industries Association, *EIA / TIA Standard: Electrical Performance for Television Transmission Systems*, Engineering Department, Series EIA/TIA-250-C, February 1990.

[8] *Military specification, Microcircuits, Linear Video A/D Converters*, MIL-M-38510 rev 5, Rome Air Development Center, Griffis AFB, New York, March 1983.

[9] DYNAD Project, "Methods and draft standards for the dynamic characterization and testing of analogue to digital converters", Project SMT4-CT98-2214, Version 3.4, July 2001.

[10] "Dynamic performance testing of A to D converters", Product Note HP-5180A, 1982.

D. Dallet and J. Machado da Silva, (eds.), Dynamic Characterisation of
Analogue-to-Digital Converters, 265–277.
© 2005 *Springer. Printed in the Netherlands.*

[11] M. Abramowitz, and I.A. Stegun, *Handbook of mathematical functions*, Dover publications Inc., New York, 1970.

[12] G. Ahn, H. Choi, S. Lim, S. Lee, C. Lee, "A 12-b, 10-Mhz, 250-mW CMOS A/D Converter", IEEE Journal of Solid-State Circuits, vol. 31, no. 12, pp. 2030-2035, December 1996.

[13] F. A. C. Alegria, and A. Cruz Serra, "Variance of the cumulative histogram of ADCs due to frequency errors", IEEE Transactions on Instrumentation and Measurement, vol. 52, no. 1, pp. 69-74, February 2003.

[14] P. Arpaia, A. Cruz Serra, P. Daponte, and C. Monteiro, "ADC testing based on IEEE 1057-94 standard - some critical notes", In Proceedings of 17th IEEE Instrumentation and Measurement Technology Conference, IMTC2000, pp. 119-124, Baltimore, MD, USA, 2000.

[15] F. Attivissimo, N. Giaquinto, and I. Kale, "INL Reconstruction of A/D Converters via Parametric Spectral Estimation", IEEE Transactions on Instrumentation and Measurement, vol. 53, no. 4, pp. 940-946, August 2004.

[16] S. Au, and B. Leung, "A 1.95-V, 0.34-mW, 12-b Sigma-Delta Modulator Stabilized by Local Feedback Loops", IEEE Journal of Solid-State Circuits, vol. 32, no. 3, pp. 321-328, March 1997.

[17] R. Baird, and T. Fiez, "A Low Oversampling Ratio 14-b 500 kHz DS ADC with a Self-Calibrated Multibit DAC", IEEE Journal of Solid-State Circuits, vol. 31, no. 3, pp. 312-320, March 1996.

[18] M. Benkais, *Méthodologie de caractérisation des circuits de conversion de données : application aux convertisseurs analogiques-numériques a facteur de mérite élevé. Mise en oeuvre dans le systeme CANTEST*. These de doctorat, 12 May 1993.

[19] M. Benkais, S. Le Masson, and P. Marchegay, "A/D converter characterization by spectral analysis in "Dual-Tone" mode", IEEE Transactions on Instrumentation and Measurement, vol. 44, no. 5, pp. 940-944, October 1995.

[20] D. Bernel, and T. Hofner, "Dynamic Parameters Describe High-Speed ADC Performance", Microwaves and RF, pp. 81-86, June 1997.

[21] M. Bertocco, C. Narduzzi, P. Paglierani, and D. Petri, "Accuracy of Effective Bit Estimation Methods", IEEE Transactions on Instrumentation and Measurement, vol. 46, pp. 1011-1015, August 1997.

[22] J. J. Blair, "A Method for Characterizing Waveform Recorder Errors Using the Power Spectral Distribution", IEEE Transactions on Instrumentation and Measurement, vol. 43, no. 5, October 1992.

[23] J. J. Blair, "Histogram measurement of ADC nonlinearities using sine wave", IEEE Transactions on Instrumentation and Measurement, vol. 43, pp. 373-383, June 1994.

[24] J. J. Blair, "Error estimates for frequency responses calculated from time-domain measurements", IEEE Transactions on Instrumentation and Measurement, vol. 47, no. 2, pp. 345-353, April 1998.

[25] J. J. Blair, "Selecting test frequencies for two-tone phase plane analysis of ADCs", IEEE Transactions on Instrumentation and Measurement, vol. 51, no. 5, pp. 976-979, October 2002.

[26] B. E. Peetz, "Dynamic Testing of Waveform Recorders", IEEETransactions on Instrumentation and Measurement, vol. 32, pp. 12-17, March 1983.

[27] P. Carbone, and D. Petri, "Quasi-coherent Equivalent-time Sampling", In *Proceedings of the 2nd International Workshop on Advanced Mathematical Models in Electrical and Electronic Measurements*, Villa Olmo, Como, Italy, July 1998.

[28] P. Carbone, and G. Chiorboli, "ADC sinewave histogram testing with quasi-coherent sampling", IEEE Transactions on Instrumentation and Measurement, vol. 50, no. 4, August 2001.

[29] P. Carbone, E. Nunzi, and D. Petri, "Statistical efficiency of the ADC sinewave histogram test", IEEE Transactions on Instrumentation and Measurement, vol. 51, no. 4, pp. 849-852, August 2002.

[30] P. Carbone, E. Nunzi, and D. Petri, "Windows for ADC Dynamic Testing via Frequency-Domain Analysis", IEEE Transaction on Instrumentation and Measurement, vol. 52, no. 6, pp. 1571-1576, December 2001.

[31] A. B. Carlson, *Communication Systems*, Mc Graw-Hill, New York, 1986.

[32] F. Chen, and B. Leung, "A High Resolution Multibit Sigma-Delta Modulator with Industrial Level Averaging", IEEE Journal of Solid-State Circuits, vol. 30, no. 4, pp. 453-460, April 1995.

[33] G. Chiorboli, M. Fontanili and C. Morandi, "A new method for estimating the aperture uncertainty of A/D converters", IEEE Transactions on Instrumentation and Measurement, vol. 47, no. 1, pp. 61-64, February 1998.

[34] G. Chiorboli, and M. Fontanili, "Cross-Correlation Noise Measurements in A/D converters", IEEE Transactions on Instrumentation and Measurement, vol. 48, no. 6, pp. 1282-1286, December 1999.

[35] T. Cho, and P. Gray, "A 10 b, 20 Msample/s, 35 mW Pipeline A/D Converter", IEEE Journal of Solid-State Circuits, vol. 30, no. 3, pp. 166-172, March 1995.

[36] D. Choy, R. Pierson, F. Trafton, B. Sheahan, V. Gopinathan, G. Mayfield, I. Ranmuthu, S. Venkatraman, V. Pawar, O. Lee, W. Giolma, W. Krenik, W. Abbott, and K. Jonhson, "An Analog Front-End Signal Processor for a 64 Mbits/s PRML Hard-Disk Drive Channel", IEEE Journal of Solid-State Circuits, vol. 29, no. 12, pp. 1596-1605, December 1994.

[37] D. Cline, and P. Gray, "A Power Optimized 13-b 5 Msamples/s Pipelined Analog-to-Digital Converter in 1.2 μm CMOS", IEEE Journal of Solid-State Circuits, vol. 31, no. 3, pp. 294-303, March 1996.

[38] C. F. Coombs, *Electronic Instrument Handbook*, McGraw-Hill, 2nd ed., 1994.

[39] G. D. Cormack, and J. O. Binder, "The Extended Function Fast Fourier Transform (EF-FFT)", IEEE Transactions on Instrumentation and Measurement, vol. 38, pp. 730-735, June 1989.

[40] G. D. Cormack, D. A. Blair, and J. N. McMullin, "Enhanced spectral resolution FFT for step-like signals", IEEE Transactions on Instrumentation and Measurement, vol. 40, no. 1, pp. 34-36, February 1991.

[41] T. Dabóczi, and I. Kollár, "Multiparameter optimization of inverse filtering algorithms", IEEE Transactions on Instrumentation and Measurement, vol. 45, no. 2, pp. 417-421, April 1996.

[42] T. Dabóczi, "Uncertainty of signal reconstruction in the case of jittery and noisy measurements", IEEE Transactions on Instrumentation and Measurement, vol. 47, no. 5, pp. 1062-1066, October 1998.

[43] I. De Lotto, and S. Osnaghi, "On random signal quantization and its effects in nuclear physics measurements", Nuclear Instrumentation Methods, vol. 56, pp. 157-159, August 1967.

[44] J. E. Dennis, and R.B. Shnabel, *Numerical Methods for Unconstrained Optimization and Nonlinear Equations*, SIAM, Philadelphia, 1996.

[45] R. De Vries, and A. J. E. M. Janssen, "Decreasing the Sensitivity of ADC Test Parameters by Means of Wobbling", Journal of Electronic Testing: Theory and Applications, vol. 15, pp. 23-29, 1999.

[46] J. P. Deyst, and T. M. Souders, "Bounds on frequency response estimates derived from uncertain step response data", IEEE Transactions on Instrumentation and Measurement, vol. 45, no. 2, pp. 378-383, April 1996.

[47] J. P. Deyst, N. G. Paulter, T. Dabóczi, G. N. Stenbakken, and T. M. Souders, "A Fast-Pulse Oscilloscope Calibration System", IEEE Transactions on Instrumentation and Measurement, vol. 47, no. 5, pp. 1037-1041, October 1998.

[48] J. Doernberg, H. S. Lee, and D. A. Hodges, "Full-speed testing of A/D converters", IEEE Journal of Solid-State Circuits, vol. 19, no. 12, pp. 820-827, December 1984.

[49] D. Duff, "ADC Achieves 12-b Resolution at 25.6 MSamples/s", Microwaves and RF, pp. 138-143, July 1991.

[50] M. O. Felix, "Differential Phase and Gain Measurements in Digitized Video Signals", SMPTE Journal, vol. 85, pp. 76-79, February 1976.

[51] E. Nunzi, P. Carbone, and D. Petri, "A Procedure for Highly Reproducible Measurements of ADC Spectral Parameters", IEEE Transactions on Instrumentation and Measurement, vol. 52, no. 4, pp. 1279-1283, August 2003.

[52] M. Flynn, and D. Allstot, "CMOS Folding A/D Converters with Current-Mode Interpolation", IEEE Journal of Solid-State Circuits, vol. 31, no. 9, pp. 1248-1257, September 1996.

[53] E. Fossum, "Digital Camera System on a Chip", IEEE Micro, pp. 8-15, May-June 1998.

[54] I. Fujimori, K. Koyama, D. Trager, F. Tam, and L. Longo, "A 5-V Single-Chip Delta-Sigma Audio A/D Converter with 111 dB Dynamic Range", IEEE Journal of Solid-State Circuits, vol. 32, no. 3, pp. 329-336, March 1997.

[55] W. L. Gans, and N. S. Nahman, "Continous and discrete Fourier transform of steplike waveforms", IEEE Transactions on Instrumentation and Measurement, vol. 31, pp. 97-101, June 1982.

[56] W. L. Gans, "The measurement and deconvolution of time jitter in equivalent-time waveform samplers", IEEE Transactions on Instrumentation and Measurement, vol. 32, pp. 126-133, March 1983.

[57] W. L. Gans, "Calibration and error analysis of a picosecond pulse waveform measurement system at NBS", Proceedings of the IEEE, vol. 74, pp. 86-90, January 1986.

[58] W. L. Gans, "Dynamic calibration of waveform recorders and oscilloscopes using pulse standards", IEEE Transactions on Instrumentation and Measurement, vol. 39, pp. 952-957, December 1990.

[59] M. F. Toner, and G. W. Roberts, "A BIST Scheme for an SNR test of a sigma-delta ADC", Proceedings of the International Test Conference, pp. 805-814, 1993.

[60] J. D. Giacomini, "High-Performance ADCs Require Dynamic Testing", Application Note AD-02, National Semiconductor, http://www.national.com/apnotes/, August 1992.

[61] N. Giaquinto, and A. Trotta, "Fast and Accurate ADC Testing Via an Enhanced Sine Wave Fitting Algoritm", IEEE Transactions on Instrumentation and Measurement, vol. 46, pp. 1020-1025, August 1997.

[62] J. Nocedal, and S. J. Wright, *Numerical Optimization*, Springer, 1999.

[63] K. levenberg, "A Method for the Solution of Certain Problem in Least Squares", Quart. Appl. Math., vol. 2, pp. 164-168, 1944.

[64] D. Marquardt, "An algorithm for Least Squares Estimation of Nonlinear Parameters", SIAM Journal on Applied Mathematics, vol. 11, pp. 431-441, 1963.

[65] G. H. Golub, and V. Pereyra, "The Differentiation of Pseudo-Inverse and Nonlinear Least Squares Problems whose Variables Separate", SIAM Journal on Numerical Analysis, vol. 10, pp. 413-432, April 1973.

[66] F. Goodenough, "ADCs Become Application Specific", Electronic Design, pp. 42-50, April 1st 1993.

[67] F. Goodenough, "Geophysicists Get 24-Bit, 0.1-W ADC To Sense Seismic Signals", Technology Advances, Electronic Design, pp. 34-48, February 20th 1992.

[68] T. Grandke, "Interpolation algorithms for Discrete Fourier Transfom weighted signals", IEEE Transactions on Instrumentation and Measurement, vol. 32, no. 2, pp. 350-355, June 1983.

[69] M. B. Grove, "Measuring frequency response and effective bits using digital signal processing techniques", Hewlett-Packard Jornal, pp. 29-35, February 1992.

[70] G. H. Hardy, and E. M. Wright, *An Introduction to the Theory of Numbers*, Oxford University Press, Oxford, 1971.

[71] F. J. Harris, "On the use of windows for harmonic analysis with the discrete Fourier Transform", Proceedings of the IEEE, vol. 66, pp. 51-83, January 1978.

[72] K. W. Hejn, and R. C. S. Morling, "A Semifixed Frequency Method for Evaluating the Effective Resolution of A/D Converters", IEEE Transactions on Instrumentation and Measurement, vol. 41, no. 2, pp. 212-217, April 1992.

[73] K. Hejn, and A. Pacut, "Generalized Model of the Quantization Error — An Unified Approach", IEEE Transactions on Instrumentation and Measurement, vol. 45, no. 1, February 1996.

[74] J. Horn, "A Basic Guide to the AC Specifications of ADCs – 2", http://www.hit.bme.hu/people/papay/edu/DSP/ACspec/Part2.htm, 2000.

[75] J. Horn, "The relationship between harmonic distortion and integral nonlinearity", http://www.hit.bme.hu/people/papay/edu/DSP/inl.htm, 2000.

[76] M. Inerfield, W. Skones, S. Nelson, D. Ching, P. Cheng, and C. Wong, "High Dynamic Range InP HBT Delta-Sigma Analog-to-Digital Converters", IEEE Journal of Solid-State Circuits, vol. 38, no. 9, September 2003.

[77] F. H. Irons, K. J. Riley, D. M. Hummels, and G. A. Friel, "The Noise Power Ratio - Theory and ADC Testing", IEEE Transactions on Instrumentation and Measurement, vol. 49, no. 3, pp. 659-665, June 2000.

[78] M. Ito, T. Miki, S. Hosotani, T. Kumamoto, Y. Yamashita, M. Kijima, T. Okuda, and K. Okada, "A 10 bit 20 MS/s 3 V Supply CMOS A/D Converter", IEEE Journal of Solid-State Circuits, vol. 29, no. 12, pp. 1531-1536, December 1994.

[79] R. C. Jaeger, "Tutorial: Analog Data Acquisition Technology - Part II", IEEE Micro, August 1982.

[80] J. F. Kaiser, and R. W. Schafer, "On the use of the I_0-sinh window for spectrum analysis'", IEEE Transactions on Acoustics, Speech, Signal Processing, vol. 28, no. 1, pp. 105-107, February 1980.

[81] G. Kang, M. Choe, and M. Song, "A High Performance CMOS A/D Converter for a Digital Camcorder System", IEEE Transactions on Consumer Electronics, vol. 42, no. 3, pp. 285-289, August 1996.

[82] D. Kim, J. Park, S. Kim, D. Jeong, and W. Kim, "A Single Chip D-S ADC with a Built-In Variable Gain Stage and DAC with a Charge Integrating Subconverter for a 5 V 9600-b/s Modem", IEEE Journal of Solid-State Circuits, vol. 30, no. 8, pp. 940-943, August 1995.

[83] K. Kim, N. Kusayanagi, and A. Abidi, "A 10-b, 100-MS/s CMOS A/D Converter", IEEE Journal of Solid-State Circuits, vol. 32, no. 3, pp. 302-311, March 1997.

[84] K. Kotani, T. Shibata, and T. Ohmi, "CMOS Charge-Transfer Preamplifier for Offset-Fluctuation Cancellation in Low-Power A/D Converters", IEEE Journal of Solid-State Circuits, vol. 33, no. 5, pp. 762-769, May 1998.

[85] N. Kusayanagi, T. Choi, M. Hiwatashi, M. Segami, Y, Akasaka, and T. Wakabayashi, "A 25 Ms/s 8-b-10 Ms/s 10-b CMOS Data Acquisition IC for Digital Storage Oscilloscopes", IEEE Journal of Solid-State Circuits, vol. 33, no. 3, pp. 492-496, March 1998.

[86] S. Kwak, B. Song, and K. Bacrania, "A 15-b, 5-Msample/s Low-Spurious CMOS ADC", IEEE Journal of Solid-State Circuits, vol. 32, no. 12, pp. 1866-1875, December 1997.

[87] C. Kyriakis, "Fundamental and Technological Limitations of Immersive Audio Systems", Proceedings of the IEEE, vol. 86, no. 5, May 1998.

[88] Y. Langard, J. L. Balat, and J. Durand, "An improved method of ADC jitter measurement", Proceedings International Test Conference, ITC94, pp. 763-770, 1994.

[89] J. Lee, H. Liu, and H. Samueli, "A Digital Adaptive Beamforming QAM Demodulator IC for High Bit-Rate Wireless Communications", IEEE Journal of Solid-State Circuits, vol. 33, no. 3, pp. 367-370, March 1998.

[90] W. J. LeVeque, *Fundamentals of Number Theory*, Dover Publications, New York, 1996.

[91] T. E. Linnenbrink, "Effective Bits: Is That All There Is?", IEEE Transactions on Instrumentation and Measurement, vol. 33, pp. 184-187, September 1984.

[92] M. Mahoney, *DSP Based Testing*, The Computer Society Press, 1987.

[93] P. Malcovati, C. Leme, P. O'Leary, F. Maloberti, and H. Baltes, "Smart Sensor Interface with A/D Conversion and Programmable Calibration", IEEE Journal of Solid-State Circuits, vol. 29, no. 8, pp. 963-966, August 1994.

[94] L. Maliniak, "CMOS Data Converters Usher In High Performance At Low Cost", Analog Outlook, Electronic Design, pp. 69-76, February 23th 1998.

[95] R. Martins, *Caracterização Dinâmica de Conversores Analógico-digital*, PhD Thesys (in Portuguese), Instituto Superior Técnico, January 1999.

[96] Y. Matsuya, and J. Yamada, "1 V Power Supply, Low-Power Consumption A/D Conversion Technique with Swing-Supression Noise Shaping", IEEE Journal of Solid-State Circuits, vol. 29, no. 12, pp. 1524-1530, December 1994.

[97] P. Maulik, N. Bavel, K. Albright, and X. Gong, "An Analog/Digital Interface for Cellular Telephony", IEEE Journal of Solid-State Circuits, vol. 30, no. 3, pp. 201-209, March 1995.

[98] M. Mayes, and S. Chin, "A 200 mW, 1 Msample/s, 16-b Pipelined A/D Converter with On-Chip 32-b Microcontroller", IEEE Journal of Solid-State Circuits, vol. 31, no. 12, pp. 1862-1872, December 1996.

[99] D. McCartney, A. Sherry, J. O'Dowd, and P. Hickey, "A Low-Noise Low-Drift Transducer ADC", IEEE Journal of Solid-State Circuits, vol. 32, no. 7, pp. 959-967, July 1997.

[100] L. G. Melkonian, "Dynamic specifications for Sampling A/D converters", Application note AN-769, National Semiconductor, http://www.national.com/apnotes/, May 1991.

[101] H. Mendonça, *Teste Dinâmico de Conversores Analógico/Digitais - Novos Métodos de Cálculo de Parâmetros de Caracterização Funcional*, PhD Thesys (in Portuguse), Faculdade de Engenharia da Universidade do Porto, September 2003.

[102] D. Mercer, "A 14-b 2.5 MSPS Pipelined ADC with On-Chip EPROM", IEEE Journal of Solid-State Circuits, vol. 31, no. 1, pp. 70-76, January 1998.

[103] J. A. Mielke, "Frequency Domain Testing of ADCs", IEEE Design and Test of Computers, pp. 64, Spring 1996.

[104] T. Miki, H. Kouno, T. Kumamoto, Y. Kinoshita, T. Igarashi, and K. Okada, "A 10-b 50 MS/s 500 mW A/D Converter Using a Differential-Voltage Subconverter", IEEE Journal of Solid-State Circuits, vol. 29, no. 4, pp. 516-522, April 1994.

[105] A. M. Mood, and F. A. Graybill, *Introduction to the Theory of Statistics*, Mc Graw-Hill, New York, 1963.

[106] C. Morandi, and L. Niccolai, "An improved code density test for the dynamic characterization of flash A/D converters", IEEE Transactions on Instrumentation and Measurement, vol. 43, no. 3, pp. 384-388, June 1994.

[107] J. Murphy, "Development of High Performance Analog-to-Digital Converters for Defense Applications", IEEE 19th Annual Gallium Arsenide

Integrated Circuit (GaAs IC) Symposium, Technical Digest, pp. 83-86, October 1997.

[108] K. Nagaraj, H. Fetterman, J. Anidjar, S. Lewis, and R. Renninger, "A 250-mW, 8-b, 52-Msamples/s Parallel- Pipelined A/D Converter with Reduced Number of Amplifiers", IEEE Journal of Solid-State Circuits, vol. 32, no. 3, pp. 312-320, March 1997.

[109] K. Nakamura, M. Hotta, L. Carley, and D. Allstot, "An 85 mW, 10 b, 40 Msample/s CMOS Parallel-Pipelined ADC", IEEE Journal of Solid-State Circuits, vol. 30, no. 3, pp. 173-183, March 1995.

[110] B. Nauta, and A. Venes, "A 70-MS/s 110-mW 8-b CMOS Folding and Interpolating A/D Converter", IEEE Journal of Solid-State Circuits, vol. 30, no. 12, pp. 1302-1308, December 1995.

[111] A. H. Nuttall, "Some windows with very good side lobe behavior", IEEE Transactions on Acoustics, Speech and Signal Processing, vol. 29, no. 1, pp. 84-91, February 1981.

[112] O. Nys, and R. Henderson, "A 19-bit Low-Power Multibit Sigma-Delta ADC Based on Data Weighted Averaging", IEEE Journal of Solid-State Circuits, vol. 32, no. 7, pp. 933-942, July 1997.

[113] F. Paillardet, and P. Robert, "A 3.3 V 6 bits 60 MHz CMOS Dual ADC", IEEE Journal of Solid-State Circuits, vol. 41, no. 3, pp. 880-883, August 1995.

[114] A. Papoulis, *Probability, Random Variables, and Stochastic Processes*, McGraw-Hill, New York, 1964.

[115] M. Pelgrom, A. Rens, M. Vertregt, and M. Dijkstra, "A 25-Ms/s 8-bit CMOS A/D Converter for Embedded Application", IEEE Journal of Solid-State Circuits, vol. 29, no. 8, pp. 879-886, August 1994.

[116] D. Petri, "Frequency-Domain Testing of Waveform Digitizers", IEEE Transactions on Instrumentation and Measurement, vol. 51, no. 3, pp. 445-453, June 2002.

[117] R. Philpott, R. Kertis, R. Richetta, T. Schmerbeck, and D. Schulte, "A 7 Mbyte/s (65 MHz), Mixed-Signal, Magnetic Recording Channel DSP Using Partial Response Signaling with Maximum Likelihood Detection", IEEE Journal of Solid-State Circuits, vol. 29, no. 3, pp. 177-184, March 1994.

[118] R. Pintelon, and J. Schoukens, "An Improved Sine-wave Fitting Procedure for Characterizing Data Acquisition Channels", IEEE Transactions

on Instrumentation and Measurement, vol. 45, no. 2, pp. 588-593, April 1996.

[119] K. Poulton, K. Knudsen, J. Corcoran, K. Wang, R. Nubling, R. Pierson, M. Chang, P. Asbeck, and R. Huang, "A 6-b, 4 GSa/s GaAs HBT ADC", IEEE Journal of Solid-State Circuits, vol. 30, no. 10, pp. 1109-1118, October 1995.

[120] W. H. Press, and S. A. Teukolsky and W.T. Vetterling and B.P. Flannery, *Numerical recipes in C: the art of scientific computing*, Cambridge University Press, 1992.

[121] J. B. Rettig, and L. Dobos, "Picosecond time interval measurements", IEEE Transactions on Instrumentation and Measurement, vol. 44, no. 2, pp. 284-287, April 1995.

[122] H. Reyhani, and P. Quinlan, "A 5 V, 6-b, 80 MS/s BiCMOS Flash ADC", IEEE Journal of Solid-State Circuits, vol. 29, no. 8, pp. 873-878, August 1994.

[123] T. Ritoniemi, E. Pajarre, S. Ingalsuo, T. Husu, V. Eerola, and T. Saramaki,"A Stereo Audio Sigma-Delta A/D-Converter", IEEE Journal of Solid-State Circuits, vol. 29, no. 12, pp. 1514-1523, December 1994.

[124] R. Roovers, and M. Steyaert, "A 175 Ms/s, 6 b, 160 mW, 3.3 V CMOS A/D Converter", IEEE Journal of Solid-State Circuits, vol. 31, no. 7, pp. 938-951, July 1996.

[125] J. C. Rudell, J. A. Weldon, J.-J. Ou, L. Lin, and P. Gray, "An Integrated GSM/DECT Receiver: Design Specifications", UCB Electronics Research Laboratory Memorandum — UCB/ERL M97/82, April 1988.

[126] M. Sauerwald, "Designing with High-Speed Analog-to-Digital Converters", National Semiconductor, Application note AD-01, May 1988.

[127] B. Schweber, "CCDs inspire ADCs to embrace signal processing", EDN Magazine, pp. 55-67, March 13th 1998.

[128] A. Cruz Serra, M. F. da Silva, P. Ramos, L. Michaeli, and J. Sǎliga, "Fast ADC Testing by Spectral and Histogram Analysis", Proceedings of the 21st IEEE Instrumentation and Measurement Technology Conference – IMTC'04, pp. 823-828, May 2004.

[129] J. Sevenhams, and Z. Chang, "A/D and D/A Conversion for Telecommunication", IEEE Circuits and Devices, pp. 32-42, January 1998.

[130] M. Shinagawa, Y. Akazawa, and T. Wakimoto, "Jitter Analysis of High-Speed Sampling Systems", IEEE Journal of Solid-State Circuits, vol. 25, no. 1, pp. 220-224, February 1990.

[131] J. Schoukens, "A Critical Note on Histogram Testing of Data Acquisition Channels", IEEETransactions on Instrumentation and Measurement, vol. 44, pp. 860-863, August 1995.

[132] T. Shu, B. Song, and K. Bacrania, "A 13-bit 10Msample/s ADC Digitally Calibrated with Oversampling Delta-Sigma Converter", IEEE Journal of Solid-State Circuits, vol. 30, no. 4, pp. 443-452, April 1995.

[133] T. Shu, K. Bacrania, and R. Gokhale, "A 10-b 40-Msample/s BiCMOS A/D Converter", IEEE Journal of Solid-State Circuits, vol. 31, no. 10, pp. 1507-1510, October 1996.

[134] D. Slepian, "Prolate Spheroidal Wave Functions, Fourier Analysis, and Unertainty-V: The Discrete Case", The Bell System Technical Journal, vol. 57, no. 5, May-June 1978.

[135] W. Song, H. Choi, S. Kwak, and B. Song, "A 10-b 20-Msample/s Low-Power CMOS ADC", IEEE Journal of Solid-State Circuits, vol. 30, no. 5, pp. 514-521, May 1995.

[136] A. B. Sripad, and D. L. Snyder, "A necessary and sufficient condition for quantization errors to be uniform and white", IEEE Transactions on Acoustics, Speech and Signal Processing, vol. 25, no. 5, pp. 442-448, October 1977.

[137] T. M. Souders, and D. R. Flach, "Accurate frequency response determinations from discrete step response data", IEEE Transactions on Instrumentation and Measurement, vol. 36, no. 2, pp. 432-439, June 1987.

[138] T. M. Souders, D. R. Flach, and J. J. Blair, "Step and frequency response testing of waveform recorders", In Proceedings of the 7th IEEE Instrumentation and Measurement Technology Conference, IMTC'90, pp. 214-220, 1990.

[139] S. Tsukamoto, I. Dedic, T. Endo, K. Kikuta, K. Goto, and O. Kobayashi, "A CMOS 6-b, 200 MSample/s, 3 V-Supply A/D Converter for a PRML Read Channel LSI", IEEE Journal of Solid-State Circuits, vol. 31, no. 11, pp. 1831-1836, November 1996.

[140] G. Uehara, and P. Gray, A 100 MHz A/D Interface for PRML Magnetic Disk Read Channels, IEEE Journal of Solid-State Circuits, vol. 29, no. 12, pp. 1606-1613, December 1994.

[141] M. Vanden Boosche, J. Schoukens, and J. Renneboog, "Dynamic testing and diagnostics of A/D converters", IEEE Transactions on Circuits and Systems, vol. 33, pp. 775-785, August 1986.

[142] R. Van De Plassche, *Integrated Analog-to-Digital and Digital-to-Analog Converters*, Kluwer Academic Publishers, ch. 5, pp. 189-203, 1994.

[143] J. Verspecht, "Compensation of timing jitter-induced distortion of sampled waveforms", IEEE Transactions on Instrumentation and Measurement, vol. 43, no. 5, pp. 726-732, October 1994.

[144] P. Vorenkamp, and R. Roovers, "A 12-b, 60-MSample/s Cascaded Folding and Interpolating ADC", IEEE Journal of Solid-State Circuits, vol. 32, no. 12, pp. 1876-1886, December 1997.

[145] X. Wang, P. J. Hurst, and S. H. Lewis, "A 12-bit 20-MS/s Pipelined ADC with Nested Digital Background Calibration", IEEE Journal of Solid-State Circuits, vol. 39 , no. 11, pp. 1799-1808, November 2004.

[146] B. Widrow, "Statistical analysis of amplitude quantised sampled data systems", IEEE Trans. AIEE, Part II Appl. Ind., vol. 79, no. 52, pp. 555-568, January 1961.

[147] B. Widrow, I. Kollár, and M.C. Liu, "Statistical Theory of Quantization", IEEE Transactions on Instrumentation and Measurement, vol. 45, no. 2, pp. 353-361, April 1996.

[148] M. Yotsuyanagi, H. Hasegawa, M. Yamaguchi, M. Ishida, and K. Sone, "A 2 V, 10 b, 20 Msample/s, Mixed-Mode Subranging CMOS A/D Converter", IEEE Journal of Solid-State Circuits, vol. 30, no. 12, pp. 1533-1537, December 1995.

[149] J. Q. Zhang, Z. Xinmin, H. Xiao, and S. Jinwei, "Sinewave Fit Algorithm Based on Total Least-Squares Method with Application to ADC Effective Bits Measurement", IEEE Transactions on Instrumentation and Measurement, vol. 46, pp. 1026-1030, August 1997.

[150] A. I. Zverev, *Handbook of filter synthesis*, John Wiley and Sons, 1967.

[151] Analog Devices, "AD7896 - 2.7 V to 5.5 V, 12-Bit, 8 us ADC in 8-Lead SOIC/PDIP, serial output 12-bit ADC", Data Sheet, Rev. C, 2003.

[152] Burr-Brown, "ADC614 - 14-Bit, 5.12 MHz Sampling Analog-To-Digital Converter", Data Sheet.

[153] Texas Instruments, "ICL7135C – 4 1/2-Digit Precision Analog-to-Digital Converters", Data sheet, September 2003.

Index